Transforming Remote Sensing Data into Information and Applications

Steering Committee on Space Applications and Commercialization

Space Studies Board
Division on Engineering and Physical Sciences
and
Ocean Studies Board
Division on Earth and Life Studies

National Research Council

NATIONAL ACADEMY PRESS
Washington, D.C.

NOTICE: The project that is the subject of this report was approved by the Governing Board of the National Research Council, whose members are drawn from the councils of the National Academy of Sciences, the National Academy of Engineering, and the Institute of Medicine. The members of the committee responsible for the report were chosen for their special competences and with regard for appropriate balance.

Support for this project was provided by National Aeronautics and Space Administration Contract No. NASW-96013, National Oceanic and Atmospheric Administration Contract No. 50-DKNA-6-90040, Stennis Space Center Order No. NS-7426, Environmental Protection Agency Grant No. X-82821401, Department of Transportation Order No. DTRS56-00-P-70077, U.S. Geological Survey Cooperative Agreement No. 00HQAG0204, and Department of the Army Order No. DACA89-99-M-0147. Any opinions, findings, conclusions, or recommendations expressed in this material are those of the authors and do not necessarily reflect the views of the sponsors.

Cover images, left to right: (1) Landsat 7 image of coral reefs at Plantantion Key, Florida, formed from buildups of calcium carbonate deposited from the skeletal remains of anthozoan cnidarians; (2) view of Santa Barbara, California, toward the northeast (with the Goleta Valley in the foreground and snow-capped Mount Abel along the skyline), generated from the Shuttle Radar Topography Mission and an enhanced Landsat satellite image; and (3) composite of two images from the Spaceborne Imaging Radar C/X-Band synthetic aperture radar flown on the space shuttle, showing two large ocean eddies at the northeast edge of the sea ice pack in the Weddell Sea, off Antarctica. SOURCE: See <http://visibleearth.nasa.gov/cgi-bin>.

International Standard Book Number 0-309-08271-4

Copies of this report are available free of charge from:

Space Studies Board
National Research Council
2101 Constitution Avenue, NW
Washington, DC 20418

Printed in the United States of America

THE NATIONAL ACADEMIES

National Academy of Sciences
National Academy of Engineering
Institute of Medicine
National Research Council

The **National Academy of Sciences** is a private, nonprofit, self-perpetuating society of distinguished scholars engaged in scientific and engineering research, dedicated to the furtherance of science and technology and to their use for the general welfare. Upon the authority of the charter granted to it by the Congress in 1863, the Academy has a mandate that requires it to advise the federal government on scientific and technical matters. Dr. Bruce M. Alberts is president of the National Academy of Sciences.

The **National Academy of Engineering** was established in 1964, under the charter of the National Academy of Sciences, as a parallel organization of outstanding engineers. It is autonomous in its administration and in the selection of its members, sharing with the National Academy of Sciences the responsibility for advising the federal government. The National Academy of Engineering also sponsors engineering programs aimed at meeting national needs, encourages education and research, and recognizes the superior achievements of engineers. Dr. Wm. A. Wulf is president of the National Academy of Engineering.

The **Institute of Medicine** was established in 1970 by the National Academy of Sciences to secure the services of eminent members of appropriate professions in the examination of policy matters pertaining to the health of the public. The Institute acts under the responsibility given to the National Academy of Sciences by its congressional charter to be an adviser to the federal government and, upon its own initiative, to identify issues of medical care, research, and education. Dr. Kenneth I. Shine is president of the Institute of Medicine.

The **National Research Council** was organized by the National Academy of Sciences in 1916 to associate the broad community of science and technology with the Academy's purposes of furthering knowledge and advising the federal government. Functioning in accordance with general policies determined by the Academy, the Council has become the principal operating agency of both the National Academy of Sciences and the National Academy of Engineering in providing services to the government, the public, and the scientific and engineering communities. The Council is administered jointly by both Academies and the Institute of Medicine. Dr. Bruce M. Alberts and Dr. Wm. A. Wulf are chairman and vice chairman, respectively, of the National Research Council.

STEERING COMMITTEE ON SPACE APPLICATIONS AND COMMERCIALIZATION

ROBERTA BALSTAD MILLER, Columbia University, *Chair*
MARK R. ABBOTT,* Oregon State University
LAWRENCE W. HARDING, JR., Horn Point Laboratory
JOHN R. JENSEN, University of South Carolina
CHRIS J. JOHANNSEN, Purdue University
MOLLY MACAULEY, Resources for the Future
JOHN S. MacDONALD, Institute for Pacific Ocean Science and Technology
JAY S. PEARLMAN, TRW, Inc.

Staff

PAMELA L. WHITNEY, Study Director
DAN WALKER, Senior Program Officer
JULIE ESANU, Program Officer
KIRSTEN ARMSTRONG, Research Assistant
CARMELA J. CHAMBERLAIN, Senior Project Assistant

---

*Until April 2001.

SPACE STUDIES BOARD

Preface

For several years, the Space Studies Board has recognized the significance of both the acceleration of opportunities for practical and operational applications of Earth observations from space and the changing economic and commercial environment for the production and management of remote sensing images. There has long been an interest throughout the remote sensing community in the development of applications. Over the past several years the combination of technological advances in remote sensing, the development of compatible geographic information software, the increased availability of data at usable scales, and greater diversity in data sources and infrastructure support have made widespread and diverse applications feasible in a broad variety of new sectors. During this same period, changes have been taking place in the roles played by data producers and consumers in the public and private sectors, the universities, and the value-adding community. Changes in the economic and policy environment for remote sensing, the growth of a commercial remote sensing industry, the expansion and proliferation of data sources worldwide, and the greater breadth of remote sensing data provided by federal agencies are the result of a number of interacting market, policy, and budgetary opportunities and incentives.

To gather data on and explore the implications of these significant changes in the environment for the production and use of remote sensing images, the Space Studies Board initiated a series of three workshops to focus on three broad areas: (1) the extension of remote sensing technologies and products into operational applications through technology transfer; (2) the conduct of scientific research in the new and evolving remote sensing environment; and (3) the development and use of remote sensing applications in the public sector, specifically state and local government. The board sought and obtained sponsorship for the series from several government agencies: the National Aeronautics and Space Administration (Headquarters and Stennis Space Center), National Oceanic and Atmospheric Administration (National Ocean Service and National Environmental Satellite Data and Information Service), Environmental Protection Agency, Department of Transportation, U.S. Army Corps of Engineers, and U.S. Geological Survey.

The first workshop, entitled "Moving Remote Sensing from Research to Applications: Case Studies of the Knowledge Transfer Process," was held at the National Academies' building in Washington, D.C., on May 3-4, 2000. The workshop included participants from federal agencies and not-for-profit organizations, academic researchers, state and local government representatives, and private sector representatives. Recognizing that the coastal zone is a particularly vital environmental arena that would benefit from enhanced remote sensing applications, the Space Studies Board collaborated with the Ocean Studies Board to develop a project whose purpose was to do the following:

- Illustrate how ground-based and in situ data collected for monitoring and assessment can be augmented by remote sensing data and images;
- Illustrate how an area understood in a research and scientific context can be developed for an applications focus;
- Introduce the coastal engineering and marine science community to existing or future remote sensing data applicable to challenges faced by the coastal engineering and marine science community (e.g., problems related to hypoxia, water quality, sediment transport, and other issues);
- Call attention to science policy issues related to the increased emphasis on the commercialization and applications of remote sensing data; and
- Evaluate the efficacy of the workshop format in bridging the gap between remote sensing technology and potential remote sensing user communities.

To prepare this report, the Steering Committee on Space Applications and Commercialization (Appendix A) drew on information from several sources: the workshop itself, including information presented by plenary speakers, in splinter sessions, and in case study presentations (Appendix B gives the workshop agenda); information presented at the planning meeting held by the steering committee in December 1999 and attended by representatives of the various sponsoring agencies and other interested individuals (Appendix C gives the agenda); and steering committee research and deliberations during and after the workshop. Not intended as a formal study of technology transfer in the context of remote sensing applications, this report reflects input received from and discussions held with a broad spectrum of actual and potential users of applications of remote sensing data,[1] with a special focus on the coastal zone. The steering committee drew on this material to prepare findings and recommendations aimed at improving the process of moving remote sensing research to operational applications.

In assessing its charge, the steering committee quickly realized that one item, to "introduce the coastal engineering and marine science community to existing or future remote sensing data applicable to challenges faced by the coastal engineering and marine science community," not only would entail an attempt to conduct technology transfer during the workshop, but also would detract from discussions about critical elements of technology transfer in the development of a broader spectrum of remote sensing applications. The steering committee thus chose to focus on extant cases, selected in collaboration with the Ocean Studies Board and experts in coastal science and engineering, of the application of remote sensing to problems in coastal science and engineering rather than on the use of remote sensing data to create new applications.

This report is directed at a diverse and growing audience, including an array of federal agencies and the broad remote sensing applications community, which comprises public and private sector providers of data, value-adding service providers, current and potential users of remote sensing data, scientists and engineers whose work spans the continuum from science to applications, and policy makers in federal, state, and local government. A similarly broad audience will be addressed by the steering committee's two remaining reports in this series.

[1]A concerted effort was made to include a balance of government, academic, not-for-profit, and private sector participants at the workshop. However, only a small number of private sector representatives attended, perhaps due to organizational schedules and demands. The steering committee is taking steps to increase private sector participation in the two additional workshops it will conduct.

Acknowledgment of Reviewers

This report has been reviewed in draft form by individuals chosen for their diverse perspectives and technical expertise, in accordance with the procedures approved by the NRC's Report Review Committee. The purpose of this independent review is to provide candid and critical comments that will assist the institution in making its published report as sound as possible and to ensure that the report meets institutional standards for objectivity, evidence, and responsiveness to the study charge. The review comments and draft manuscript remain confidential to protect the integrity of the deliberative process. We wish to thank the following individuals for their review of the report:

Peter Banks, XR Ventures, LLC,
John A. Harrington, Jr., Kansas State University,
Robert C. Harriss, National Center for Atmospheric Research,
Christopher T. Hill, George Mason University,
Jean-Michel M. Rendu, Newmont Mining Corporation (retired),
Walter Schmidt, Florida Geological Survey, and
James Yoder, University of Rhode Island.

Although the reviewers listed above have provided many constructive comments and suggestions, they were not asked to endorse the conclusions or recommendations, nor did they see the final draft of the report before its release. The review of this report was overseen by John V. Evans, Comsat Corporation (retired). Appointed by the National Research Council, he was responsible for making certain that an independent examination of this report was carried out in accordance with institutional procedures and that all review comments were carefully considered. Responsibility for the final content of this report rests entirely with the authoring committee and the institution.

Contents

EXECUTIVE SUMMARY 1

1 REALIZING THE POTENTIAL OF REMOTE SENSING 7
Background, 8
Turning Remote Sensing Data into Information, 13
Bridging the Knowledge Gap, 16
Moving from Research to Applications, 19

2 MEETING THE INFORMATION NEEDS OF END USERS IN THE COASTAL ZONE 23
Unique Capabilities, and Some Current Limitations, of Remote Sensing, 24
Meeting User Requirements: Observations Based on the Case Studies, 25
Improving the Prospects for Remote Sensing Technology Transfer, 29

3 RESPONDING TO ISSUES CRITICAL TO THE DEVELOPMENT OF SUCCESSFUL APPLICATIONS 34
Cost-Effectiveness of Applications, 34
Timeliness of Data, 38
Reliability and Continuity of Data, 40
Data Formats and Standards, 41
Education, Training, and the Remote Sensing Workforce, 42
Intellectual Property Issues, 44

4 RECOMMENDATIONS 46
Bridging Gaps, 46
Transferring Technology, 47
Findings and Recommendations, 48

APPENDIXES

A Biographical Information for Steering Committee Members and Workshop Speakers 57

B Workshop Agenda and Participants 66

C Planning Meeting Agenda 71

D Acronyms 74

Executive Summary

Over the past decade renewed interest in practical applications of Earth observations from space has coincided with and been fueled by significant improvements in the availability of remote sensing data and in their spectral and spatial resolution. In addition, advances in complementary spatial data technologies such as geographic information systems and the Global Positioning System have permitted more varied uses of the data. During the same period, the institutions that produce remote sensing data have also become more diversified. In the United States, satellite remote sensing was until recently dominated largely by federal agencies and their private sector contractors. However, private firms are increasingly playing a more prominent role, even a leadership role, in providing satellite remote sensing data, through either public-private partnerships or the establishment of commercial entities that serve both government and private sector Earth observation needs. In addition, a large number of private sector value-adding firms have been established to work with end users of the data.

These changes, some technological, some institutional, and some financial, have implications for new and continuing uses of remote sensing data. To gather data for exploring the importance of these changes and their significance for a variety of issues related to the use of remote sensing data, the Space Studies Board initiated a series of three workshops. The first, "Moving Remote Sensing from Research to Applications: Case Studies of the Knowledge Transfer Process," was held in May 2000. This report draws on data and information obtained in the workshop planning meeting with agency sponsors, information presented by workshop speakers and in splinter group discussions, and the expertise and viewpoints of the authoring Steering Committee on Space Applications and Commercialization. The recommendations are the consensus of the steering committee and not necessarily of the workshop participants.

Rather than trying to cover the full spectrum of remote sensing applications, the steering committee focused on civilian remote sensing applications in the coastal environment.[1] The workshop featured three case studies in coastal management involving (1) the application of Sea-viewing Wide-Field-of-view Sensor (SeaWiFS) data in monitoring harmful algal blooms, (2) the use of airborne lidar bathymetry for monitoring navigation channels, and (3) the use of both satellite and aerial remote sensing to identify sewage outflows. All three provided detailed information on the applications as well as problems encountered in developing them, allowing the steering committee to learn from the real-world experiences of particular users.

In addition, participants in five workshop splinter sessions—on education and training, institutional, technical, and policy issues in technology transfer, and user awareness and needs—identified and discussed more general barriers and bottlenecks that interfere with the development of remote sensing applications and also explored ways to overcome such problems. Plenary presentations focused on research on technology transfer; science and policy issues in the coastal zone; a comparison of remote sensing technology transfer with respect to geographic information systems and the Global Positioning System; and new directions in the use of remote sensing data. This material provided a basis for much of the steering committee's analysis and figured significantly in its development of the report's findings and recommendations.

BASIC OBSERVATIONS

To encourage finding more effective ways to develop new and useful applications of remote sensing data, the steering committee considered barriers to as well as opportunities for developing successful applications through the transfer of knowledge and technology.[2] Its examination of the remote sensing technol-

---

[1]Although a great deal of excellent work on operational applications has been done within the defense community, those developments were independent of civil remote sensing in terms of both budgets and technologies and hence they are not within the purview of this report.

[2]The steering committee approached technology and knowledge transfer in terms of the application of remote sensing data and images in the public, private, and not-for-profit sectors (regardless of whether they were produced by public or private sector image providers). These applications may depend on data from either the public or the private sector. Similarly, the process of technology transfer can take place within or across government agencies, between the public and the private sectors, within the private sector, and between the private or government sectors and the not-for-profit sector. At issue is not where the data originate or who uses them, but rather how to develop useful, operational applications.

ogy transfer process led to the identification of a number of gaps that must be bridged in order to develop effective civilian applications:

- The gap between the raw remote sensing data collected and the information needed by applications users. Users need information, and the process of transforming data into information is a critical step in the development of successful remote sensing applications.
- The gap in communication and understanding between those with technical experience and training and the potential new end users of the technology. Producers and technical processors of remote sensing data must be able to understand the needs, cultural context, and organizational environments of end users. Education and training can also help to ensure that new end users have a better understanding of the potential utility of the technology.
- The financial gap between the acquisition of remote sensing data and the development of a usable application. The purchase of data is only the first of a large number of steps affecting the cost of a successful application. An organization, commercial firm, or government agency that wants to incorporate remote sensing applications into its operations must be prepared for a long-term financial investment in staff, ongoing training (both technical and user training), hardware, and software, at a minimum. Alternatively, the potential user organization should be prepared to purchase these services from a value-adding provider.

Another recurring theme in workshop discussions was the need for data continuity. In light of the heavy, up-front investment required to develop and use remote sensing applications, organizations as well as individual users have to be assured of a reliable and continuous source of both data and information.

FINDINGS AND RECOMMENDATIONS

Life-Cycle Costs

Finding. The full, life-cycle cost of developing and using remote sensing data products goes beyond obtaining the data and includes, among others, staff for data processing, interpretation, and integration; education and training; hardware and software upgrades; and sustained interactions between technical personnel and end users (see Chapter 3). Although many of these costs are incurred at the time a technology is first employed, the life-cycle costs and benefits of remote sensing applications are not well understood.

Recommendation 1. NASA's Office of Earth Science, Applications Division, in consultation with other stakeholders (e.g., agencies that use remote sensing data, such as the U.S. Geological Survey, Department of Transportation, Environmental Protection Agency, and U.S. Department of Agriculture; private companies;

state and local government users; and not-for-profit institutions), should mount a study to identify and analyze the full range of short- and long-term costs and benefits of developing remote sensing applications and the full costs of their implementation by public, nongovernmental, and other noncommercial users. In addition, NASA should support economic analyses to reduce the start-up costs of developing new remote sensing applications.

Education and Training

Finding. Training is an integral component of efforts to bridge the gap between remote sensing professionals and end users (see Chapters 3 and 4). Remote sensing involves sophisticated technology, and specialized training is required to process the data, convert it into information, and interpret the results. Many agencies and organizations either lack the financial resources to provide such training or do not understand the importance of periodic retraining for technical staff.

Recommendation 2. Federal agencies such as NASA, the National Oceanic and Atmospheric Administration (NOAA), the U.S. Department of Agriculture, the U.S. Geological Survey (USGS), and others should provide the seed funding for developing remote sensing training and educational materials. Agencies should consider, as an initial step, using the Small Business Innovation Research (SBIR) program to solicit proposals for developing training materials and courses, to foster the uses of remote sensing data in applications, and to encourage commercial enterprises to provide these services.

Outreach

Finding. Reducing the social distance between application developers and end users is a means of encouraging successful technology transfer (see Chapters 2 and 3). Unless those who create applications (e.g., scientists, engineers, and technicians) and those who use them (e.g., government, not-for-profit, and private sector applied users, policy makers, and natural resource managers) understand the roles of others involved in the process, they will not be able to communicate effectively and the development of applications will suffer.

Recommendation 3. Federal agencies, including those that produce remote sensing images and those that use them, should consider creating "extern" programs with the purpose of fostering the exchange of staff among user and producer agencies for training purposes.

For example, NASA, NOAA, and USGS could create an extern program in collaboration with potential user agencies, such as the Environmental Protection Agency, the U.S. Army Corps of Engineers, the U.S. Department of Agriculture, the Department of Transportation, and others and in so doing could produce trained staff to serve as brokers for information and further training. Similar exchanges could be organized between universities and state and local governments and between commercial companies and government.

Recommendation 4. The Land Grant, Sea Grant, and Agricultural Extension programs should be expanded to include graduate fellowships and associateships to permit students to work at agencies that use remote sensing data. Such programs could help to improve communication and understanding among the scientists and engineers who develop applications for remote sensing data and the agencies that use them.

NASA's Space Grant program could be extended to include these training activities, much as the Land Grant program has fostered the development of agricultural extension agents.

Applications Research

Finding. Although many remote sensing applications emerge from basic research, the development of applications is not accorded the recognition associated with publication in scientific journals. Researchers have few professional incentives to produce applications. The research-to-applications model developed in other fields, such as pharmaceutical research and many fields of engineering, could be emulated by the Earth sciences. Yet even if this model were to be adopted in areas related to remote sensing, there are at present few funding opportunities for work that spans the divide between research and applications.

Recommendation 5. Resources, separate from funding for basic research, should be made available to federal agencies such as NASA, the National Oceanic and Atmospheric Administration, the Environmental Protection Agency, the U.S. Geological Survey, the Department of Transportation, the National Science Foundation, and others for support of research on remote sensing applications and remote sensing applications derived from basic research. In addition, these agencies should establish joint research announcements aimed at fostering the development of applications for remote sensing data through basic research.

Requirements of Applications Users

Finding. Many remote sensing applications have specific requirements, including continuity in data collection, consistency in format, frequency of observations, and access to comparable data over time. It is important that the requirements of those who use applications are communicated to both public and private sector data producers throughout the process of designing new technologies and producing and disseminating remote sensing data.

Recommendation 6. Both public and private sector data providers should develop mechanisms to obtain regular advice and feedback on applications requirements for use in their planning processes. Advisory bodies that are consulted for input to these decisions should routinely include applications users.

Recommendation 7. Data preservation should be addressed by all data providers as a routine part of the data production process to ensure continuity of the data record and to avoid inadvertent loss of usable data.

Standards and Protocols

Finding. The lack of standard data formats, open and available protocols, and standard validation and verification information inhibits the spread of remote sensing applications (see Chapter 3).

Recommendation 8. The use of internationally recognized formats, standards, and protocols should be encouraged for remote sensing data and information. The work of the OpenGIS Consortium and the Federal Geographic Data Committee serves as an important international and national coordinating mechanism for efforts in standards development that should be continued.

These and other entities pursuing common remote sensing data formats and standards should consult with the sensor and software vendors to ensure that data acquired from the use of new technologies for data acquisition, analysis, and storage and distribution are consistent with other data sets.

Utility of Workshop Format

Finding. In general, the workshop as a mechanism for gathering data provided the steering committee with the information and insight it needed to understand issues related to technology transfer and remote sensing applications and to make recommendations about more effective ways to foster the development of applications.

In retrospect, as outlined in Chapter 4, the steering committee recognizes several strengths, and some areas for improvement, in the use of a workshop format.

1

Realizing the Potential of Remote Sensing

Over the past decade, a renewed and expanding interest in practical applications of Earth observations from space and airborne platforms has coincided with and been fueled by changes in the data, in how they can be used, and in who produces them. There have been significant improvements in the availability of remote sensing data and in their spectral and spatial resolution. In addition, the data can be adapted for more varied uses because of the extension and advancement of complementary spatial data technologies, such as geographic information systems and the Global Positioning System, which can be used in conjunction with remote sensing data. During the same period, the institutional support for producing remote sensing data has also become more diversified. In the United States, satellite remote sensing was until recently dominated by federal agencies and their private sector contractors and was focused on reconnaissance, scientific and technological innovation, and operational weather monitoring and prediction. Although the private sector has long been actively involved in providing airborne images for a variety of applied needs, commercial companies have only recently begun to provide satellite remote sensing imagery. Increasingly, private sector firms are playing a more central role, even a leadership role, in providing satellite remote sensing data, either through public-private partnerships or through the establishment of commercial entities that serve both government and private sector Earth observation needs. Public sector organizations and private firms also provide technology for instrumentation flown on airborne platforms that enable the development of additional products and services. In addition, many private sector value-adding firms have been established to work with end users of the data.

These changes, some technological, some institutional, and some financial, have implications for new and continuing uses of remote sensing data. Recognizing the importance of these changes and their significance for a variety of issues related to the use of remote sensing data, the Space Studies Board established the Steering Committee on Space Applications and Commercialization to hold a series of three workshops to explore three sets of related issues.[1] Based on the first workshop, titled "Moving Remote Sensing From Research to Applications: Case Studies of the Knowledge Transfer Process" and held in May 2000, this report focuses on the process of technology and knowledge transfer in transforming remote sensing data into applications.

BACKGROUND

Remote sensing has long been recognized as a means of obtaining data and information to meet perceived needs. Systematic remote sensing began in the period between World Wars I and II with aerial photography used for military reconnaissance and photogrammetry. The Cold War emphasis on collecting intelligence to monitor the U.S.-Soviet arms race stimulated the rapid technological advancement of satellite remote sensing capabilities for military applications. During the same period progress was made in the development of remote sensing technologies applicable to civilian needs: Box 1.1 lists milestones over the past four decades.[2] TIROS 1, the first meteorological satellite, was launched in 1960, and the Earth Resources Technology Satellite series, later renamed Landsat, began operating in 1972. With Landsat came a continuous source of satellite images that could be used routinely for a variety of civilian applications such as mineral and oil exploration, crop monitoring, and natural resource management. Early programs such as the Large Area Crop Inventory Experiment, a program to forecast the yield of specific crops, called attention to the possibilities for developing a variety of useful applications of remote sensing data, although the effort never achieved the ambitious goals set by its proponents.[3]

By the 1980s, as a result of budgetary problems and a declining interest in

---

[1]The second workshop will address the conduct of scientific research in the new and evolving remote sensing environment, and the third will focus on public sector applications of remote sensing data.

[2]Although the military sector has a long history of collecting remotely sensed data and transforming it into information for military uses, this report focuses on the development of civilian applications. Both in the allocation of budgetary resources and in the use of the technology, military developments evolved separately from civil remote sensing, and defense and military systems were designed to prevent access to the data rather than to encourage their widespread use.

[3]Pamela E. Mack, Viewing the Earth: The Social Construction of the Landsat Satellite System. Cambridge, Mass.: MIT Press, 1990, pp. 146-158.

civilian applications at NASA, which had taken the technical lead in civilian Earth observation, responsibility for civil operational sensors and remote sensing satellites had been transferred to the National Oceanic and Atmospheric Administration (NOAA).[4] NASA concentrated its energies on developing sensors for gathering scientific data,[5] and resources for Earth observation were directed increasingly to instruments intended primarily to meet scientific and environmental data needs. In the late 1980s, the agency committed itself to building the Earth Observing System (EOS), an environmental satellite system that was defined in terms of both NASA and interagency environmental observation needs.

As NASA was realigning its activities to meet scientific and policy data needs, interest in the role of the private sector in Earth observation was growing. The French SPOT system (Système pour l' Observation de la Terre) was created through a public-private partnership with the specific goal of selling data and developing a self-sustaining commercial enterprise.[6] SPOT Image was organized to operate the system and sell data; the French government provided support for the spacecraft system. During the same period, the U.S. government experimented with commercialization of the Landsat system, awarding a contract to the Earth Observation Satellite Company (EOSAT) in 1985 to operate the Landsat system and sell the data on the commercial market.[7] This experiment, however, did not meet expectations.[8]

NASA moved back into the arena of remote sensing applications with the establishment of the Earth Observations Commercialization/Applications Program (EOCAP) in 1993 and a university-government-industry collaborative pro-

---

[4]Presidential Directive 54 (PD 54) of November 1979 assigned NOAA responsibility for all civil land remote sensing satellite systems.

[5]Office of Technology Assessment, U.S. Congress, *The Future of Remote Sensing from Space: Civilian Satellite Systems and Applications,* Washington, D.C., Government Printing Office, 1993, p. 49; Space Studies Board, National Research Council, *Earth Observations from Space: History, Promise, and Reality,* Washington, D.C., National Academy Press, 1995, p. 113.

[6]Office of Technology Assessment, U.S. Congress, *The Future of Remote Sensing from Space: Civilian Satellite Systems and Applications,* Washington, D.C., Government Printing Office, 1993, p. 53.

[7]Space Studies Board, National Research Council, *Earth Observations from Space: History, Promise, and Reality,* Washington, D.C., National Academy Press, 1995, p. 114; Office of Technology Assessment, U.S. Congress, *The Future of Remote Sensing from Space: Civilian Satellite Systems and Applications,* Washington, D.C., Government Printing Office, 1993, p. 49.

[8]Prices increased significantly during the period of transition from a government to a private operator. The higher prices resulted in lower sales, and consequently the use of the data and the development of applications decreased. Landsat 6, which was to have been operated by EOSAT, failed to reach orbit, further curtailing EOSAT's ability to seek a profit from Landsat data on the commercial market. The U.S. government did not obtain a commercial operator for Landsat 7 and transferred the program back to government control.

BOX 1.1
Milestones in Civil Satellite Remote Sensing

1960 The first meteorological satellite, TIROS 1, takes first images to be used in weather forecasts.

1960 CORONA, a reconnaissance satellite, takes its first image. CORONA imagery was declassified in 1995.

1968 Apollo 8 returns the first pictures of Earth from deep space.

1972 NASA launches the first civilian Earth resource satellite, Earth Resources Technology Satellite (later named Landsat 1).

1975 The Large Area Crop Inventory Experiment becomes the first Landsat applications program.

1975 NASA launches Landsat 2 and 3.

1979 NOAA assumes control of the Landsat program from NASA pursuant to executive order.

1982 NASA launches Landsat 4.

1983 NOAA transfers the nation's civil, operational remote sensing satellites to the private sector.

1984 Land Remote Sensing Commercialization Act establishes the process for the commercialization of land remote sensing satellites.

1985 NASA launches Landsat 5.

1985 NOAA transfers the Landsat program to the Earth Observation Satellite Company (EOSAT), a private operator.

1985 Nimbus-7 Total Ozone Mapper confirms ozone depletion.

1986 France launches the first commercially oriented Earth observation satellite, SPOT 1.

1986 The media uses Landsat 5 data to monitor the Chernobyl nuclear power plant disaster, marking the first widespread use of satellite imagery in the news media.

1991	NASA establishes the Earth Observing System (EOS) program in response to a U.S. Presidential initiative to provide in-depth scientific understanding about the functioning of the Earth as a system.
1991	Landsat 5 monitors burning oil wells and other environmental effects of the Gulf War.
1992	Land Remote Sensing Policy Act of 1992 transfers the Landsat program back to the government and provides for the continuation of the program with Landsat 7.
1993	Department of Commerce issues the first license to operate a private remote sensing system to EarthWatch for the Early Bird satellite.
1993	EOSAT's launch of Landsat 6 fails.
1995	U.S. delegate to the United Nations shows satellite images of mass graves in Bosnia to the UN Security Council.
1995	Russian firm, Sovinformsputnik, sells 2-m imagery.
1995	Canada launches the first operational synthetic aperture radar satellite, Radarsat 1.
1996	NASA creates lead center for commercial remote sensing applications development.
1997	Congress and the Office of Management and Budget direct NASA to acquire Earth science data products from commercial sources.
1998	NASA's Office of Earth Science creates the Applications Division.
1998	Department of Defense awards its first contract to procure high-resolution satellite images gathered by privately owned satellites.
1999	NASA launches Landsat 7.
1999	Space Imaging, Inc. launches the first commercial 1-m resolution satellite, IKONOS 2.
1999	NASA launches the first Earth Observing System (EOS) satellite, Terra.
2000	Department of Commerce issues first licenses to operate remote sensing systems to collect 0.5-m imagery.

gram of affiliated research centers around the same time.[9] NASA's Stennis Space Center was given responsibility for developing commercial remote sensing applications and administering NASA's Science Data Buy, a demonstration program encouraging NASA to purchase scientific research data from the private sector. Through EOCAP, NASA established partnerships with private companies and business alliances seeking to develop and market remote sensing applications products. According to NASA, its role is to share in the financial risks and provide technical guidance to industry during the development, validation, and dissemination of prototypes for commercially viable applications. The affiliated research centers program funds university efforts to develop and test new uses of remote sensing data and data analysis tools in partnership with both small and large private sector companies. Regardless of their outcome, NASA's programs to encourage the development of viable commercial uses of remote sensing data exemplify the changes under way in the existing relationships among federal science agencies, the private sector in the form of the commercial remote sensing industry, and both scientific and applied users of remote sensing data.

In 1998, NASA established the Applications Division to foster and expand the uses of Earth Science Enterprise (ESE) products in the public and private sectors. The formation of the Applications Division, parallel in structure to the ESE Research Division, signaled the significance accorded to remote sensing applications within NASA.

Today scientific and operational remote sensing images produced by U.S. government agencies, aerial and high-resolution satellite images from the private sector, and remote sensing images from international sources contribute to the growing abundance of Earth observation data. In 1999 alone, there were three major launches of civil Earth observation satellites: Landsat 7, IKONOS, and Terra, one of which (IKONOS) was launched by a private sector firm, Space Imaging, Inc. Moreover, the expected launches of new satellites over the next several years with capabilities for gathering data of even higher spatial and spectral resolution in both the public and private sectors will add to the rich array of possibilities for additional applications. (One forecast projects some 15 commercial land imaging satellites to be launched during the period from 2001 to 2006.[10])

In addition to more data suppliers, there is now a more diverse infrastructure for producing applications, including a growing number of value-adding producers, university centers, and consultants. Advances in computing capabilities and the development and availability of geographic information technologies have given added impetus to the use of remote sensing data in new types of

---

[9]For more information on affiliated research centers see <http://www.crsp.ssc.nasa.gov/scripts/arc/>, accessed on March 2, 2001.

[10]Stoney, William E., "Summary of Land Imaging Satellites (with Better Than 30 Meters Resolution) Planned to Be Operational by 2006," McLean, Va., Mitretek Systems, March 20, 2001.

applications. Socioeconomic, epidemiological, and ecosystem databases can be integrated with remote sensing data in a geographic information system to improve understanding of complex spatial and environmental relationships. The Internet, in particular, the World Wide Web, provides a means for scientists and applications users to identify existing remote sensing data and imagery and to obtain them rapidly. The intersection of these various technological advances offers the potential for a new period in the application of remote sensing to public policy, governance, and commercial needs. For example, the use of remotely sensed imagery in the media made such events as the Chernobyl disaster, oil wells set afire during the Persian Gulf War, and atrocities in Bosnia and Kosovo visible to the general public in ways that were previously not possible.[11]

TURNING REMOTE SENSING DATA INTO INFORMATION

Space-based remote sensing provides a new source of information that cannot be easily obtained in other ways and that promises both economic and social benefits. To fulfill this promise will require a better understanding of cost-effective ways to realize potential useful applications. Workshop discussions involving remote sensing applications users made it clear that the utility of the technology will come not from the data itself but rather from the information that can be derived from the data. These users emphasized that turning data into useful information is central to technology transfer and the development of successful applications.

Workshop discussions suggested that to date new applications of remote sensing data have been developed largely by individuals or organizations that already possessed both the necessary technical expertise and an understanding of potential uses of the data. Participants pointed out that remote sensing data can initially appear complicated and possibly even irrelevant to potential end users who make policy and management decisions. Such users need easily understood information that can be used to address economic, social, environmental, and other policy questions. For this reason, research to enable interpretation of the data, and transformation of remote sensing data into usable information, are critical steps in the development of applications.

To enjoy widespread use, remote sensing data must be made accessible to information consumers who may not have the technical expertise currently required to use such data. Past approaches to applications development have

[11]There has also been a long history of civilian applications of classified imagery. For example, MEDEA, an environmental scientific advisory group, has shown the advantages of using intelligence data and imagery for civil environmental applications. See Pace, Scott, O'Connell, Kevin M., and Lachman, Beth E., *Using Intelligence Data for Environmental Needs: Balancing National Interests*, Santa Monica, Calif., RAND, 1997.

BOX 1.2
Limitations of Common Approaches to Applications Development

Models that motivated many of the previous government programs to develop successful applications of remote sensing may have inhibited their success. Despite government investment in demonstration programs, subsidies to data producers and data users, and applications partnerships between and within the public and private sectors aimed at developing economically viable applications of remote sensing, the full potential of the technology for routine or operational applications has not yet been realized. In particular, these previous models for applications development do not address several of the challenges noted in the workshop: (1) the knowledge and communications gap between technical experts and information consumers, (2) the lack of incentives for many end users to adopt remote sensing applications, and (3) the need for applications users to anticipate and meet the full spectrum of costs associated with the development and use of remote sensing applications.

Demonstration research programs are often premised on the belief that successful applications in both the public and private sectors will emerge from research projects that have demonstrated new uses of remote sensing data. The federal role in this approach has been to identify areas of need or promise and to provide financial support for university-led research projects in these areas. However, another necessary step, moving from research results to operational demonstrations of utility that will persuade nontechnical end users to adopt remote sensing applications, is often not taken. This is the need addressed, for example, by the county extension agent (now known as a county extension educator) in agricultural technology transfer. A key component of the success of this process is the "spannable social distance" between the county agent and his clients and, as a result, the ability of the county agent to understand and speak directly to farmers' needs and priorities.[1] The county agent will meet with both farmers and seed industry representatives to determine the best way to present the leading hybrid varieties to other farmers. A demonstration plot of all varieties sold in the county will usually be planted along a well-traveled roadway so that other farmers can observe them throughout the growing season. Yield data are obtained and summary tables of this information are provided at the end of the growing season so that farmers can use this information in selecting seed for the following year.

Subsidies are used in two ways. One, as illustrated by NASA's Earth Observa-

stopped short of realizing remote sensing's full potential (Box 1.2). The demand for applications will be driven by requirements for information rather than by the technical capabilities of the end users. Unlike those who developed the first applications of remote sensing, many new applications users are likely to have little, if any, knowledge of remote sensing technology or how it is employed to derive information. They will be concerned instead with the accuracy and timeliness of the data and with its relevance for specific tasks and decisions.

Potential users of remote sensing information want to be assured that its

tions Commercialization/Applications Program (EOCAP), involves providing an incentive to the private sector to invest in remote sensing applications by creating a partnership in which the government serves as a financial partner. EOCAP projects have involved applications for agriculture, geology, urban land use, real estate management, and telecommunications. A second approach involves subsidizing the use of data for applications, perhaps by providing data at no cost through government data programs or through such vehicles as the experimental science data buy program that gives scientists access to free commercial data through a competitive proposal process.[2] To the extent that scientific research leads to specific applications, workshop discussions suggested, the direct subsidization of scientific data acquisition through the public and private sectors may promote applications. However, this approach is not oriented to bridging the gap between data and information. More specifically, it does not address the issue of end user information needs.

Finally, a *partnership approach* has been employed in the development of SPOT in France and Radarsat in Canada. The central government serves as the financial partner of a private sector firm in supporting the construction, launch, and operation of remote sensing satellites for both scientific and applied uses. The European Space Agency's decision to have ERS-1 and Envisat synthetic aperture radar data distributed on a commercial basis is another example of the public-private partnership approach.[3] In this case, governments subsidize the production of data but have not become involved in the process of converting data to information, and have not attempted to address the full range of applications costs (see Chapter 4) that were discussed in the workshop.

Each of these three approaches emphasizes the role of government financial support in the development of new applications, but none addresses user needs or the institutional, workforce, legal, and other nontechnical issues that arise in the successful development of applications.

---

[1]Although the county agent and the farmer share a common language, education, and background, the county agent has more education and training, is knowledgeable about research, and serves as a link between the farmer and the knowledge base in state universities and federal agencies.

[2]The science data buy and its implications for Earth science research will be explored in some detail in a subsequent report by the steering committee.

[3]See "ESA Hands Radar Satellite Responsibility to Industry," Space News, Vol. 11, No. 11, October 30, 2000, pp. 1, 34.

value will surpass the institutional investments involved in acquiring and using the information. In a commercial market, there must be a balance between the value of the information, as perceived by end users, and the revenue necessary to support the information delivery system. In the public sector, the value of the information must be weighed against alternative uses of the funds needed to support the work of transforming data into information. Achieving the needed balance depends on both the intrinsic information content of the raw data produced by a remote sensing instrument and the way the data are processed to produce new information.

In the case studies examined at the workshop, it was clear that a critical element in producing information of value is processing, which involves two steps: preprocessing and the conversion of data to information. Preprocessing turns raw data into accurately calibrated measures of precisely located physical variables such as reflectance, emittance, temperature, and backscatter. The knowledge base underlying this step is usually well developed, although research may be required for developing specific applications (as in the case, for example, of developing algorithms for using SeaWiFS data to monitor *Gymnodinium breve*, a species associated with harmful "red tides"; see Chapter 2, Box 2.2).

However, the scientific knowledge base to support the conversion of data to information is far less developed. Transforming technical data into a form that is meaningful to nontechnical users—a process often including either the integration of remote sensing data with other types of data or scientific research to characterize the data (or both)—is highly dependent on the information requirements of applied users and on the skills of technical experts. For example, a digital elevation model that was extracted from remotely sensed light detection and ranging (lidar) data provided increased accuracy over conventional digital elevation models and thus proved valuable in helping to determine optimum routes for new Norfolk and Southern railroad lines.[12] In another example, a private sector firm transformed satellite imagery into maps tailored for the specific needs of commercial and sports fishermen, showing where albacore rather than swordfish were likely to be present.

The diversity of end users' information needs that might be met by the same initial set of physical variables is depicted in Figure 1.1, which illustrates several simultaneous data-to-information conversion processes.

BRIDGING THE KNOWLEDGE GAP

Data must be transformed to information and knowledge if the goal of developing successful operational applications of remote sensing data is to be met.[13] On one side of the gap are the scientists, engineers, and technologists who construct and operate instruments to measure parameters in the Earth system using spacecraft and aircraft. On the other side are actual and potential end users (the

---

[12]Cowen, D.C., Jensen, J.R., Hendrix, C., Hodgson, M.E., and S.R. Schill, "A GIS-Assisted Rail Construction Econometric Model That Incorporates LIDAR Data," *Photogrammetric Engineering and Remote Sensing* 66(11): 1323-1328, 2000.

[13]Such a transformation of data to information was demonstrated in the case studies presented at the workshop (see Chapter 2).

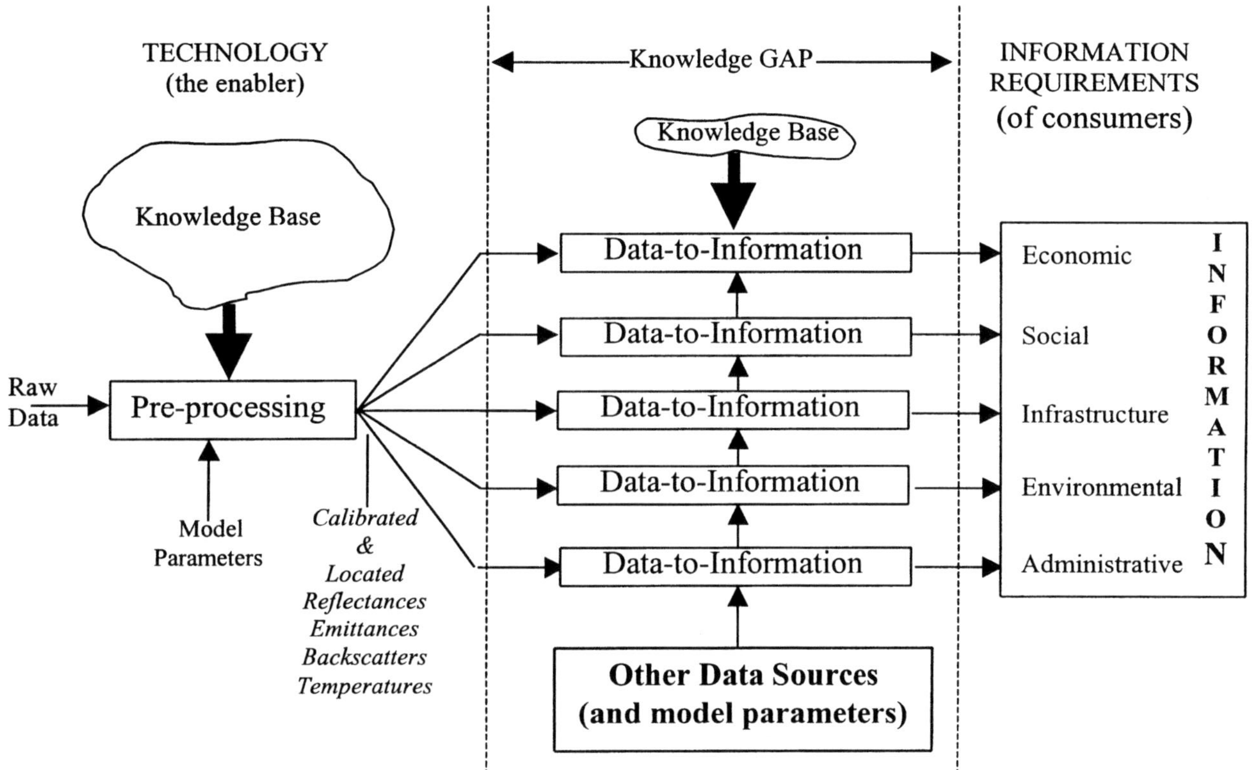

FIGURE 1.1 Processing data into information.

information consumers) who have information requirements but know little about using remote sensing technology to satisfy them and, critically, have little if any financial, personal, or institutional motivation to consider such an approach.[14] The presence within an organization of a highly motivated person or "champion" who can get an organization to recognize the potential benefits of using remote sensing data and applications can be critical to overcoming gaps in communication.

Nevertheless, those who develop sensors, collect and analyze data, and develop products to address scientific or technical questions often have few opportunities to communicate with prospective users who lack essential technical expertise. This lack of communication constitutes a significant barrier to technology transfer. Bridging the knowledge gap will depend not only on improved communication among technologists and users (see Box 1.3), but also on research focused on converting data into information.

A key barrier to transforming data and images into meaningful information is the limited understanding of how to convert measurements made from space into information of ecological, economic, social, infrastructure, environmental, or administrative value. As workshop presentations pointed out, improving this knowledge base requires the involvement of those who are knowledgeable about the physics of remote sensing and the technologies that support it. In addition, both social and natural scientists could provide and integrate complementary data for use in the transformation of raw remote sensing data into usable information, and researchers could also refine or extend the utility of an existing remote sensing application.

However, the current model of conducting scientific research and publishing the results in peer-reviewed journals that emphasize original new work can discourage the pursuit of research on applications, given that many scientific disciplines and universities often judge the merit of a researcher or a project by publication alone. Moreover, funding organizations generally issue grants for original research rather than work on applications, creating another disincentive to turning research into applications.

In workshop discussions, participants suggested the need for an alternative to peer-reviewed publication as the end product of basic research. In some fields, such as pharmacology and engineering, the end product of a research project is often an application. Mechanisms such as grants and NASA Research Announcements that encourage the development of applications research would benefit the

[14]For example, in one case study discussed in the workshop, sanitation authorities resisted using remote sensing to improve their capabilities for monitoring coastal sewage discharge, possibly because of the potential for positive signals to complicate existing monitoring and compliance issues (see Chapter 2, Box 2.3). Other case studies and examples discussed in the workshop and planning meeting highlighted the importance of creating incentives for end users to adopt remote sensing applications.

practical use of Earth science data. In addition, some new journals are publishing peer-reviewed research focusing on scientific research and its use in decision making and applications.[15] But there is little support for research to explore the linkages between basic research and applied remote sensing and the opportunities those linkages offer for developing new applications. An exception is land grant colleges and universities, which encourage the practical application of research to local or state issues.

MOVING FROM RESEARCH TO APPLICATIONS

Today, new mechanisms are needed for transferring the results of federally funded research to applications that will benefit operations within and beyond federal agencies.

Technology transfer, at the heart of which is the creation of useful knowledge and information, is a critical element in the development of new applications of remote sensing data that is not always well understood. Rather than being confined to cost considerations that tend to focus on the initial expense of acquiring imagery, and to privatization that involves the transfer or licensing of technology to private firms, the process of remote sensing technology and knowledge transfer is better discussed, as the steering committee learned, in a broader context that includes policy and institutional issues, new users' requirements, education and training, and technical issues.

Applications of remote sensing data and images for public, private, and not-for-profit uses, produced by public or private sector providers, may be developed from data originating in either the public or the private sector. Similarly, technology transfer can take place within or across government agencies, between the public and the private sectors, within the private sector, and between the private or government sectors and the not-for-profit sector. At issue is not where the data originate or who uses them, but rather how to develop useful, operational applications.

Despite the growing number of users who have taken advantage of the opportunities to apply remote sensing data to practical problems—in, for example, coastal management, monitoring environmental change, mapping, natural resource management, and public health—an even larger set of users could potentially benefit from the data. As became evident in the workshop, however, extending the benefits of remote sensing to potential new users is a complex process of technology and knowledge transfer that goes far beyond the initial task of creating market-based incentives to purchase data. Remote sensing special-

---

[15]See *Ecological Applications*, available online at <http://esapubs.org/esapubs/journals/applications_main.htm> accessed October 10, 2001.

BOX 1.3
Technology and Knowledge Transfer

The process of converting data into information is a form of technology and knowledge transfer. There has been a wealth of research on technology transfer over the past several decades, as well as a growing body of experience in technology transfer and research utilization programs,[1] only a small slice of which was presented at the workshop. This research indicates an emerging consensus not only about what technology transfer is, but also about what it is *not*. It is not generally thought to be a unidirectional, linear process that begins with basic research discoveries, which are then moved into applied research and development, and finally ends with the production and dissemination of usable applications. Nor is it a simple process of disseminating advanced technological resources to potential users. Instead, as both ongoing research and the case studies presented at the workshop make clear, the process of technology transfer is interactive and can begin at any point—with a market or user need, with applied research, with a technology or technological product, or with basic research. What is critical to this process is what David Roessner[2] calls a "spannable social distance" across each interface of components in the system.[3] By this he means that the cultural and communicative distance between producers and users of a new technology must be small.

Roessner's emphasis on spannable social distance is based on his examination of the experience of the federal government in technology transfer and research utilization programs in the 1960s and 1970s.[4] He concluded that both passive and reactive technology transfer mechanisms and programs were generally less effective than those that were active or required collaboration among producers and end users from the start.[5] Moreover, although technology transfer was more likely to be successful when applications were developed by users with substantial technical capabilities, it also required strong, ongoing personal interactions between the suppliers of the technology and the new users of that technology. Bridging the gap illustrated in Figure 1.1 will require considerable interaction between technical experts and information consumers to convert data into information of value to the end user. This interaction may also require intermediaries that are close to the cultures of both the user and the technical expert.

ists, for example, must obtain a fundamental understanding of the fit between remote sensing data and the needs of potential users. To do so may require overcoming such nontechnical barriers and bottlenecks as the lack of understanding by end users of the potential of the technology, the absence of a trained in-house technical workforce, or restrictions on the use of data or software that can effectively slow down or impede the adoption of new applications. The lack of communication among remote sensing specialists, data users, and potential infor-

[1]One recent study on the technology transfer process is Ruttan, Vernon W., *Technology, Growth, and Development: An Induced Innovative Perspective,* New York, Oxford University Press, 2001. For additional reading on the technology and knowledge transfer process see: David, Paul A., "Technology Diffusion, Public Policy, and Industrial Competitiveness" in Ralph Landau and Nathan Rosenberg, eds., *The Positive Sum Strategy: Harnessing Technology for Economic Growth,* Washington, D.C., National Academy Press, 1986; Cowan, Robin, David, Paul A., and Foray, Dominique, "The Explicit Economics of Knowledge Codification and Tacitness," *Industrial and Corporate Change: Special Issue,* Vol. 9, Issue 2, June 1, 2000; David, Paul A., "Rethinking Technology Transfers: Incentives, Institutions and Knowledge-based Industrial Development," in Charles Feinstein and Christopher Howe, eds., *Chinese Technology Transfer in the 1900s: Current Experience, Historical Problems, and International Perspectives,* Cheltenham, England, Edward Elgar, 1997; David, Paul A., and Dominique Foray, "Information Distribution and the Growth of Economically Valuable Knowledge: A Rationale for Technological Infrastructure Policies" in Morris Teubal, Dominique Foray, Moshe Justman, and Ehud Zuscovitch, eds., *Technological Infrastructure Policy: An International Perspective,* Dordrecht, Kluwer Academic Publishers, 1996.

[2]J. David Roessner, a professor of public policy at Georgia Institute of Technology, spoke on the topic "Technology Transfer Process" at the May 2000 workshop on which this report is based.

[3]For the original research on spannable social distance, see Rogers, Everette M., Eveland, J.D., and Bean, Alden S., *Extending the Agricultural Extension Model,* Institute for Communication Research, Stanford University, 1976.

[4]Roessner, J. David. "Evaluating Government Innovation Programs: Lessons from the U.S. Experience," *Research Policy* 18: 343-359, 1989.

[5]See Ballard, Steven, James, Thomas E., Adams, Timothy I., Devine, Michael D., Malysa, Lani L., and Meo, Mark, *Innovation Through Technical and Scientific Information: Government and Industry Cooperation,* Westport, Conn., Quorum Books, 1989; Doctors, S.I., *The Role of Federal Agencies in Technology Transfer,* Cambridge, Mass., MIT Press, 1969; Gruber, W.H., and Marquis, D.G., *Factors in the Transfer of Technology,* MIT Press, 1969; Havelock, R.G., *Planning for Innovation,* Ann Arbor, Mich., Center for Research on the Utilization of Scientific Knowledge, 1969; Hough, Granville, *Technology Diffusion, Federal Programs and Procedures,* Mt. Airy, Md., Lomond Books, 1975; and Rogers, Everette M., and Shoemaker, Floyd, *Communication of Innovations,* 2nd Edition, New York, Free Press, 1971.

mation consumers is one of the greatest barriers to expanding the use of remote sensing data.

Drawing on workshop discussions and material presented in the workshop's case studies, the steering committee examined these barriers and bottlenecks and developed suggestions for ways to deal with them, taking as a starting point the requirements of end users in the coastal zone for successful operational applications. To that end, the steering committee collaborated with the NRC's Ocean

Studies Board, which provided a foundation of research and access to scientific expertise for developing the workshop and the report. The use of remote sensing in the coastal zone has been discussed in a number of Ocean Studies Board and other NRC reports.[16]

In identifying barriers and bottlenecks as well as successful approaches to overcoming them, the steering committee's goal in this report is to advance the dialogue about how to succeed in the development of remote sensing applications.

[16]As early as 1992, the Ocean Studies Board called for greater use of remote sensing data in ocean and coastal research and policy and called for new partnerships with both federal remote sensing data providers and industry data providers to accomplish this effort. Recommendations regarding the application of remote sensing data in NRC reports go beyond scientific need, however, and extend into policy, monitoring, and other operational responsibilities in the coastal zone. See: Ocean Studies Board, National Research Council, *Oceanography in the Next Decade: Building New Partnerships*, Washington, D.C., National Academy Press, 1992, p. 151. In addition, other National Research Council reports have called attention to the importance of remote sensing to coastal and ocean research and policy. See: *Clean Coastal Waters: Understanding and Reducing the Effects of Nutrient Pollution* (Ocean Studies Board, 2000); *Review of NASA's Earth Science Enterprise Research Strategy for 2000-2010* (Space Studies Board, Board on Atmospheric Sciences and Climate, Board on Earth Sciences and Resources, and Ocean Studies Board, 2000); *Global Environmental Change: Research Pathways for the Next Decade* (Board on Sustainable Development, 1999); *From Monsoons to Microbes: Understanding the Ocean's Role in Human Health* (Ocean Studies Board, 1999); *Global Ocean Science: Toward an Integrated Approach* (Ocean Studies Board, 1999); and *Restoring and Protecting Marine Habitat: The Role of Engineering and Technology* (Marine Board, 1994), all published by National Academy Press, Washington, D.C.

2

Meeting the Information Needs of End Users in the Coastal Zone

Moving from demonstration projects to successful operational applications of remote sensing data requires that the desired information meet applications users' needs, which depends in turn on users and technical experts being able to discuss and understand those needs. Identifying a prospective user's requirements is essential. Often the process of developing applications for new end users, many of whom may be decision makers or policy makers in the public and private sectors, also requires research and/or the integration of multiple sources of data and information. Equally important are the conditions under which data and/or applications are made available to users, requiring that issues such as the timeliness, continuity, and stability of data; reliability of access; viability of data formats and processing; intellectual property rights; and operational cost-effectiveness be addressed satisfactorily.

Examination of the case studies presented at the workshop and summarized in this chapter offered an opportunity to draw on the experience gained through applications of remote sensing in the coastal zone.[1] The three cases selected for examination include the application of SeaWiFS data to monitor harmful algal blooms (HABs) that appear to be increasing in coastal waters,[2] the use of airborne lidar bathymetry to monitor channel and harbor status to ensure safe navi-

---

[1]A later workshop to be organized by the steering committee will explore opportunities for and barriers to the use of remote sensing data in the public sector, particularly in state and local governments.

[2]Anderson, D., *The Ecology and Oceanography of Harmful Algal Blooms (ECOHAB): A National Research Agenda.* Woods Hole, Mass., Woods Hole Oceanographic Institution, 1995, p. 66.

gation, and the use of satellite and aerial remote sensing to determine water mass trajectories and so gauge the impacts on water quality of marine sewage disposal.

UNIQUE CAPABILITIES, AND SOME CURRENT LIMITATIONS, OF REMOTE SENSING

Satellite remote sensing can significantly enhance the information available from traditional data sources because it can provide synoptic views of large portions of Earth. Satellite imagery can also expand the spatial dimensions of limited and sometimes costly field or point-source sampling efforts. Some satellite sensors cover areas that may be physically or politically inaccessible or that are too vast to survey with traditional methods. Remote sensing can also provide consistent repeat coverage at relatively frequent intervals, making detection and monitoring of change feasible. Satellite-derived data and information are also useful for applications that require fine spatial resolution such as surveys of urban and suburban land use, land cover for agricultural purposes, and natural resources; surveys for coastal management; and measurements of water quality in limnological and oceanographic applications.

The disadvantages of satellite remote sensing include the inability of many sensors to obtain data and information through cloud cover[3] (although microwave sensors can image Earth through clouds) and the relatively low spatial resolution achievable with many satellite-borne Earth remote sensing instruments. In addition, the need to correct for atmospheric absorption and scattering and for the absorption of radiation through water on the ground can make it difficult to obtain desired data and information on particular variables. Satellite remote sensing creates large quantities of data that typically require extensive processing as well as storage and analysis. Finally, data from satellite remote sensing are often costly if purchased from private vendors or value-adding resellers, and this initial cost, together with intellectual property restrictions, can limit the dissemination of products from such sources.

In many instances, there may be an advantage to combining the large-scale, synoptic data that are accessible from space with higher-resolution surveys of key locations that can be made from other platforms, such as aircraft. Aerial photography, for example, has a competitive advantage in applications that require fine spatial resolution of small areas or that involve areas subject to frequent cloud cover, especially in cases where repeat coverage is needed. Another advantage of aerial photography is that surveys can be scheduled for specific purposes and locations. Aircraft-carried instruments of other types, including visible, thermal,

---

[3]Cloud cover can be particularly problematic for optical sensors that collect imagery over the West Coast. For example, marine stratus clouds often cover the West Coast during the morning overpass of the Landsat satellite.

and microwave sensors, provide high-resolution data of many kinds and thus represent an important part of the mix of remote sensing capabilities for Earth observation. Among the disadvantages of aircraft remote sensing are the relatively limited spatial coverage that can be obtained compared to satellite instruments, the recurring expense of deploying multiple flights, weather restrictions, and lack of synopticity over large scales.

MEETING USER REQUIREMENTS: OBSERVATIONS BASED ON THE CASE STUDIES

The three case studies presented at the workshop by government agency officials and a private sector representative (Boxes 2.1, 2.2, and 2.3) outline projects developed for the application of remote sensing, describe the challenges in developing data and information, and provide examples of barriers and bottlenecks to producing useful applications. The information needs of the coastal zone managers who are the applications users in these case studies address a range of problems. The cases are not representative of all remote sensing applications developments; rather, they illustrate the types of challenges that can arise in moving from research to information and applications.

On balance, the case studies presented at the workshop suggest that successful remote sensing applications are likely to be those that draw on the unique characteristics of the data and cannot be accomplished in a cost-effective manner with other sources of data. Discussions at the workshop emphasized that it is the good correspondence between the information requirements of end users and the specific attributes of remote sensing data that will make the use of the technology advantageous. Remote sensing data that do not meet the specific information needs of end users or whose use does not fit into their accepted practices must be transformed into usable information to enable users to apply the technology.

The case studies and discussions at the workshop also highlighted the importance of easily understood demonstration projects. Potential end users were willing to participate in a demonstration project, especially when they were not required to contribute financially to the project. In turn, user feedback provided those conducting the demonstration project with valuable information on the effectiveness of the application in time to modify the demonstration before completion. Demonstration projects are one means of bridging the gap between the information needs of end users and unprocessed remote sensing data.

A common thread in the case studies and other examples of remote sensing applications in the coastal zone was the requirement to obtain consistent data and information over regular intervals on key ecosystem variables that might serve as indices of change. Remote sensing can be an attractive technology for monitoring coastal ecosystems, whose high spatial and temporal variability mean that data must be collected over large areas and for long periods to enable identification of trends. Such large-scale sampling from boats and ships alone is prohibi-

BOX 2.1
Applying SeaWiFS Data to Monitoring Harmful Algal Blooms

The Challenge

Marine planktonic algae, or phytoplankton, are essential elements of the marine environment, but some species are detrimental, forming harmful algal blooms (HABs). Toxic HAB species, such as the dinoflagellate *Gymnodinium breve (G. breve),* cause "red tides" that can harm fish and wildlife, cause illness in humans, and have a significant economic impact. Sampling HABs effectively is a difficult problem because of the large areas of ocean they may cover and the logistical costs of traditional sampling. Improved monitoring of HABs would increase understanding of the inception of HABs and the conditions that promote their growth. Rapid identification of HABs is essential for state managers to protect public health. Satellite imagery of ocean color has the potential to provide information on the distribution and abundance of HABs at high frequency and with suitable spatial and temporal resolution.

Remote Sensing Application

In partnership with the National Oceanic and Atmospheric Administration (NOAA) and the Naval Research Laboratory, the Environmental Protection Agency (EPA) Advanced Monitoring Initiative program sponsored a demonstration project to use satellite data from the Sea-viewing Wide Field-of-view Sensor (SeaWiFS) ocean color instrument to help detect outbreaks in the Gulf of Mexico of the potentially harmful *G. breve* from its optical signature. Regional SeaWiFS data on chlorophyll concentration have the potential to give an early warning of *G. breve* blooms. This program was designed to develop algorithms for using SeaWiFS data to monitor *G. breve* and to communicate the results to health officials charged with identifying health and safety issues related to changing environmental conditions.

tively expensive, except in the most confined of water bodies. In the Chesapeake Bay and in the Mississippi River plume in the Gulf of Mexico—both highly fertilized regions that are manifesting water quality problems associated with anthropogenic nutrient loading from their large watersheds—intensive monitoring programs have been developed that are being augmented effectively with remote sensing.

The case studies illustrated both barriers to developing effective applications and bottlenecks that slow or complicate the process. For example, in the Army Corps of Engineers SHOALS project, which uses airborne lidar to survey navigation channels and harbors, the long lead time from concept to application was a

Results

The demonstration program verified the usefulness of space-based SeaWiFS data for monitoring *G. breve* in the Gulf of Mexico. The correspondence between SeaWiFS data on chlorophyll concentrations[1] and cell concentrations of *G. breve* determined from ship samples suggests that satellite imagery may be useful for early detection and tracking of these harmful algal blooms (HABs). The use of SeaWiFS data to identify bloom locations and track population movements allows more extensive collection of in situ samples from appropriate locations, thus augmenting more costly and less effective sampling approaches. The program has transitioned into an operational effort: States identify a potential bloom event, NOAA's Coast Watch Program collects relevant SeaWiFS data, NOAA scientists analyze the data, and the states redirect their sampling accordingly. The program has also identified an apparent relationship between bloom events and climate. There are indications that an absence of blooms during some years and an abundance in others might be linked to large-scale weather patterns, such as El Niño and La Niña events. Improved coordination is essential to ensure that the findings of the demonstration project are transmitted to partners in the states. Efficient, cost-effective, and understandable communication of reports on HABs is integral to the success of efforts at the local level to monitor HABs.

Barriers

Barriers to the development of applications included confusion about access to and distribution of SeaWiFs data; competition among partners and institutional barriers; costs and difficulties in coordinating a multiagency, multistate project; and development of an effective means, perhaps modeled after the daily weather reports, to communicate HAB reports to a state user.

---

[1]Chlorophyll is a ubiquitous plant pigment occurring in all phytoplankton taxa, and careful verification of indications from ocean color images is essential to use this tool in predicting or monitoring HABs.

significant deterrent to developing future applications (Box 2.2). In the demonstration project using data from the ocean color sensor SeaWiFS to identify HABs, a period of scientific and technical research was required to enable the use of imagery as a tool for predicting when and where blooms might occur (Box 2.1). In both cases, the common impediment to effectively transferring remote sensing technology to prospective information users was the gap between user requirements and the technical capabilities of the data. Remote sensing data are not a "magic bullet"; they have advantages and disadvantages that will affect their utility for practical applications. Often, developing applications of value for end users will require coupling remote sensing data with other data and/or doing additional research aimed at creating new, useful data products. Remote sensing

BOX 2.2
U.S. Army Corps of Engineers SHOALS Airborne Lidar Bathymetry Program

The Challenge

The U.S. Army Corps of Engineers (USACE) is responsible for maintaining approximately 25,000 km of navigation channels and more than 600 ports and harbors. These channels must be assessed, often annually, for the movement of sand and sediment into shoals that impede safe navigation. Many channels require repeated dredging. Conventional hydrographic surveys using boats with acoustic fathometers are costly, slow, and do not meet all the USACE requirements for navigation and shore protection project monitoring.

Remote Sensing Application

USACE developed an airborne lidar system that allowed it to survey more channels without increasing its budget. The lidar technology works by transmitting green and infrared laser signals into the water. The green signal reflects off the sea bottom, and the infrared signal reflects off the water's surface; the time differential between the two signals provides the water depth. The USACE's objective was to stimulate private industry interest and investment in the technology by demonstrating the viability of lidar through the development of the Scanning Hydrographic Operations Airborne Lidar Survey (SHOALS). SHOALS was initiated in 1994. After demonstrating the technology and characterizing its capabilities, the USACE adopted the technology and SHOALS for operational use. SHOALS remains a government-owned, privately operated system.

Results

The lidar system can survey waters too shallow to allow boats to collect data and can extend the hydrographic survey onto the beach or shore. The information obtained by SHOALS at a navigation project, including ebb and flood shoals, adja-

data are usually just one element of an effective application and seldom offer a complete solution.

In the case studies described in Boxes 2.1 through 2.3, the scale of the spatial coverage provided by remote sensing is of paramount importance, facilitating detection of potentially harmful events over hundreds to thousands of square kilometers, such as the movement of sewage effluent or the development of HABs. To yield a predictive tool, however, data obtained by remote sensing must be integrated with confirming evidence gathered from other sources, such as counts of bacteria or measures of concentrations of toxins in the case of sewage effluent, and species enumeration and identification in the case of HABs. Achieving a match between the needs of users and the potential of remote sensing data to address those needs is accomplished through research and the integration of data and information from multiple sources, including remote sensing. This

cent shorelines, channels, and dredge disposal areas, exceeds what conventional surveys can produce cost-effectively and thus contributes to a more complete understanding of sediment movement in and around USACE navigation projects. SHOALS has helped to enable new business practices within the USACE that should lead to improved navigation channel maintenance. Strategic partnering by district planning, engineering, and operations has provided for regional, rather than individual, project approaches to sediment management. SHOALS has supported this new business initiative because of its ability to rapidly and cost-effectively conduct regional scale coastal surveys. In addition to improved partnerships with the districts, the SHOALS program has partnered with the U.S. Navy and the National Ocean Service of the National Oceanic and Atmospheric Administration and has conducted surveys for the U.S. Geological Survey, state governments, and several foreign countries. The program has also extended its applications to include nautical charting, shoreline mapping, coastal monitoring, coral reef mapping, and military operations.

Barriers

The initial institutional barriers to implementing SHOALS surveys in the USACE centered on issues related to acceptance of a new way of doing business. The data collectors, those using boats for hydrographic surveys, resisted because it was not their system. There were questions about survey costs and about how to compare costs of conventional versus remotely sensed hydrographic surveys. This issue was complicated by the fact that a typical conventional navigation project survey does not produce the same product as a SHOALS survey. Data users, on the other hand, were much quicker to order a SHOALS survey. However, they had problems initially because the data files were much larger than those obtained with conventional surveys, and few computer tools and models existed to allow easy utilization of the data. The usefulness of the SHOALS system is limited in areas of poor water quality and at depths over 60 meters. This led to a multiplatform approach that uses boats in areas where the lidar is not effective.

does not represent a barrier for coastal managers, who are used to working with multiple sources of data.

IMPROVING THE PROSPECTS FOR REMOTE SENSING TECHNOLOGY TRANSFER

As discussed in Chapter 1, the crux of the technology transfer process is transforming raw data from remote sensing into useful information. Discussions at the workshop and material in the case studies presented there stressed that a key element in this process is collaboration between remote sensing experts and end users to ensure a match between what the data can provide and what information is needed. Eventually the active role of technical experts in developing new applications will become routine as the applications become operational and the end users develop expertise sufficient to make use of the products.

BOX 2.3
Satellite and Aerial Remote Sensing for Monitoring Coastal Sewage Discharge

The Challenge

Coastal pollution from sewage effluent is a major problem in southern California and other U.S. coastal regions. Inadequate containment may result in beach closings and long-term pollution with significant health and economic impacts. In southern California, beach closures during the tourist season result in the loss of millions of dollars of revenue to local businesses. In some cases, cities or their sanitation districts may be fined appreciable sums for lack of compliance. Monitoring discharges is essential for U.S. coastal sanitation districts that operate under the jurisdiction of an Environmental Protection Agency (EPA) permit process. Renewal of permits is dependent on evidence of compliance derived from regular field monitoring of physical and biological properties. Most sampling is done only on a weekly or monthly basis because of the high costs of field sampling. This schedule leaves large gaps when the sanitation districts and other agencies may lack information on the trajectories of wastewater plumes.

Remote Sensing Application

Incorporating satellite and aerial remote sensing into sewage outfall monitoring programs offers several advantages: (1) large areas can be sampled synoptically at high spatial and temporal resolution for costs significantly lower than those of traditional field surveys; (2) limited field data from key locations can be significantly extended by combining them with remotely sensed information that permits spatial interpolation to broad regional coverage; and (3) gaps in coverage from field sampling can be filled using remotely sensed data of improved frequency and regularity of coverage, and these data can serve as an early-warning system for undetected sewage effluent events. Ocean Imaging Corp. organized a feasibility study with the Orange County Sanitation District to develop new commercial applications for remote sensing of temperature, ocean color, and turbidity. The study used primarily time series of sea surface temperature data from the Advanced Very High Resolution Radiometer (AVHRR) sensor to compute upper-layer currents and estimates of turbidity derived from visible AVHRR data.

Unlike skilled technical practitioners who are generally knowledgeable about remote sensing and have the expertise to analyze the data and imagery, many end users, who often have very different backgrounds and skills, are generally not equipped to interpret remote sensing data without significant technical assistance or training. In addition, many end users have little interest in remote sensing per se or in the technical characteristics of the additional information integral to applications that can help their decision making. Thus, workshop participants pointed out, education is needed to increase end users' awareness of remote sensing capabilities. In addition to understanding the broad characteristics of remotely sensed data, applications users should be encouraged to discuss their

Results

Both products of the study were found to be useful for detecting anomalous discharge-surfacing events and for helping to separate beach contamination events caused by an outfall from those caused by tidal or rain-caused discharges of stagnant water from nearby lagoons. The current analyses, which were verified by field sampling, helped estimate the possibility of effluent reaching the beaches. Follow-on projects for EPA evaluated other remote sensing data types (radar, optical, and multispectral) for monitoring new outfall and sewage runoff in San Diego and across the border with Mexico.

Barriers

The pace of remote sensing technology transfer for coastal runoff and effluent discharge applications is hampered by the following:

- *Data timeliness.* In most monitoring situations, especially when a spill or similar event is occurring, data must be processed, analyzed, and delivered to the end user in less than 24 hours to be useful for guiding management or remedial efforts. This is possible only with data from AVHRR, the Sea-viewing Wide Field-of-View Sensor, and possibly the Moderate Resolution Imaging Spectrometer, which offer poor spatial resolution for most applications.
- *Data cost.* All synthetic aperture radar and high-resolution optical remote sensing data are currently priced too high to be affordable for effective continuous coastal monitoring efforts. Given the constantly changing ocean environment, remotely sensed data lose their value extremely quickly—yesterday's image has little or no value today.
- *Resistance to increased monitoring capability.* Remote sensing cannot replace field monitoring, which is needed to sense bacteria, ammonia, or similar variables directly. Adding remote sensing to field sampling programs raises the potential for more frequent *apparent* positive signals that will only complicate meeting current monitoring, performance, and compliance requirements. There is little incentive for the districts to spend funds that will further complicate their situation.

information needs with technical personnel. A survey by NOAA's Coastal Services Center,[4] discussed at the workshop, indicated that coastal managers may not communicate their specific needs to technical personnel with sufficient clarity to permit them to develop appropriate applications of remote sensing data. Moreover, many who develop sensors, collect and analyze data, and develop products to address scientific or technical questions are unable to communicate

[4]NOAA Coastal Services Center, "Coastal Resource Management Customer Survey," Charleston, South Carolina, NOAA Coastal Services Center, 1999.

with prospective users who lack essential technical expertise. This gap in communication constitutes a significant barrier to technology transfer.

Coastal managers, who may use ocean color information from SeaWiFS to identify areas of high chlorophyll concentrations that accompany the development of harmful algal blooms, rely on information derived from the remote sensing data and made available to them by technically cognizant intermediaries in the transfer process (see Box 2.1). In this case, remote sensing data on the potential development of a HAB event, as indicated by high concentrations of chlorophyll in a satellite ocean color scene, must be coupled with additional data on the species of algae associated with the high concentration of chlorophyll, such as the data that can be obtained by analyzing in situ water samples. The remote sensing data must be turned into useful information that can be used effectively by the coastal managers to make decisions about health impacts on wildlife and humans.

Education and training are of paramount importance in linking user needs and remote sensing capabilities and in informing those with technical expertise about the day-to-day challenges faced by existing end users and potential new users. From the end users' perspective, data and information from remote sensing can be used most effectively by those with a level of expertise that includes an understanding of the more sophisticated aspects of the underlying technologies. For example, an end user who is familiar with remote sensing theory and appropriate image analysis methods is more likely to use data obtained by remote sensing than is a novice who lacks this understanding. Well-trained people are, therefore, one of the most important components of the remote sensing technology transfer process.

Increased awareness of remote sensing's potential to meet end users' needs will also stimulate the use of remote sensing applications and help to encourage the emergence of a viable market for remote sensing information and services. In the SHOALS project, for example, aircraft lidar for bathymetric surveys initially complemented more traditional shipboard approaches to gathering data, demonstrating the potential for obtaining additional information by employing remote sensing technologies to address the problem of monitoring harbor and channel depths. The role of education and training in developing remote sensing information products, knowledge, and services is addressed in Chapter 3.

Collaboration in day-to-day problem solving can also help to bridge the gap between users and technologists. In addition, value-adding companies, which employ technical experts who provide remote sensing products, information, and services and who act as intermediaries in matching a customer's needs with the company's remote sensing capabilities, can facilitate understanding of the operating and decision-making processes of these two different communities. Workshop participants also proposed "externships" as a means of fostering the exchange of staff between federal agencies that produce remote sensing data and products and the agencies that use them.

Finally, applications users usually require a stable and continuing supply of information, whether from remote sensing or other sources, and may opt for continuity of mediocre or even inadequate data rather than rely on an undependable source of superior data. They may also be deterred from obtaining new types of data if technical uncertainties associated with collecting or accessing the data appear to be high. Remote sensing data whose availability is limited are of little practical value to an end user who requires a regular stream of data to meet his or her information needs. Such issues must be addressed if remote sensing technology transfer is to succeed.

3

Responding to Issues Critical to the Development of Successful Applications

It was clear from discussions at the workshop and in the steering committee's subsequent deliberations that the acquisition of data is merely the first step in developing successful applications of remote sensing. The path from obtaining data and imagery to operationalizing an application is long and complex, involving individuals and organizations with diverse requirements and needs. To better understand the process, the steering committee considered certain concerns voiced repeatedly by workshop participants and identified a series of implementation issues as being critical to the development of successful new applications. These issues are the cost-effectiveness of applications; the timeliness, reliability, and continuity of data and data products; standardization of data formats; workforce and educational issues; and intellectual property issues. The steering committee was particularly interested in identifying barriers to new applications, bottlenecks that slowed or derailed the adoption of applications, and, most importantly, responses to these problems that can minimize or circumvent them.

COST-EFFECTIVENESS OF APPLICATIONS

Cost is a critical issue in the adoption of remote sensing applications both for those who develop an application and for its end users. Costs tend to vary by application. However, unless a remote sensing application is cost-effective and the value of the data or the resulting program efficiency or quality exceeds the associated costs over time, it will not be adopted or maintained.

Two types of costs are incurred when developing new applications: the cost of the data and the institutional costs of developing and maintaining the application. Data can be expensive. Federal program managers reported to the steering committee that the initial cost of the data is often the most important element in determining the extent to which a new measurement or monitoring method is used, even if the information derived from the new method offers significant improvements over the old. In the case discussed (the use of remote sensing data for monitoring harmful algal blooms), uncertainties about the cost of SeaWiFs data was an issue in the Environmental Protection Agency's use of remote sensing data in this application (see Chapter 2, Box 2.1). NOAA's National Ocean Services representatives also reported that the cost of commercially produced remote sensing data obtained from both satellites and aircraft makes it difficult to achieve the agency's mission. In addition, the workshop case study on coastal sewage discharge monitoring, presented by Ocean Imaging Inc., showed that the costs of data impeded effective, continuous coastal monitoring—a definite barrier in a dynamic application area that requires near-real-time data to monitor change (see Chapter 2, Box 2.3).

For the applications user, moreover, the cost of a remote sensing information product begins rather than ends with the cost of the data. For example, an organization that decides to use remote sensing imagery in-house must hire trained staff or provide technical training to current staff, acquire the computer hardware and software needed to manipulate and store the data sets, and purchase other data for integration with the remote sensing imagery. Such expenses may be in addition to costs for existing capital equipment and human resources that are not fungible. Even when these expenses have been met, workshop participants indicated the transformation of remote sensing data into usable information requires additional investments. Much of the available remote sensing software is not easy to use, and according to discussions at the workshop, interpreting and analyzing the data can require extensive experience with frequent training required to maintain skills. Consequently, the expense of creating the institutional infrastructure required for developing in-house remote sensing applications constitutes a barrier for some organizations and a temporary bottleneck for others. Finally, as suggested in Chapter 2, incorporating the new information into ongoing decision processes requires a further investment of time and effort in demonstrations, training, and lengthy discussions with nontechnical decision makers to communicate the meaning of the application and its results.

An organization beginning to use remote sensing must balance the initial investments required to obtain information from this new technology against competing (and compelling) investments of time, personnel, and financial resources for other purposes. Within any organization, the use of remote sensing applications inevitably competes with the use of other, often traditional, sources of information for decision making (provided by employees already in place). An example of the tension between existing technology and new technology is noted in the SHOALS program described in Chapter 2. Moreover, in many cases expenditures for existing information sources (personnel, infrastructure, information) must be sustained at the same time that new remote sensing expenses are assumed.

The costs associated with continued use of traditional data can appear small when compared with the costs of adding the institutional infrastructure necessary to produce information through a new technology like remote sensing. Some organizations may choose not to support and maintain the in-house infrastructure (equipment, skilled personnel, data acquisition capabilities) required to develop and use remote sensing applications. A viable alternative is for an organization to obtain remote sensing information products directly from an external service provider, often referred to as a value-adding company. Value-adding service providers handle the selection of appropriate remote sensing data, processing of the data, and development of the application, along with other services, to meet a user's needs. These companies can help clarify the information needs of end users and then act as brokers in securing the skills and services to address those needs. As with outsourced services in general, this approach might reduce initial costs, particularly those for acquisition of imagery and fixed personnel costs. It could also, however, involve searching for an appropriate service provider in an unfamiliar field and, if the demand for the applied products grows, spending large amounts in direct costs over a number of years in payments to the service provider. Moreover, as noted in previous Space Studies Board reports, any organization using remote sensing applications will benefit from a level of internal technical knowledge and experience so as to be a "smart buyer" of remote sensing products and services.[1]

Given the budgetary difficulties and at times the reluctance of organizations to support regular in-house training, the development of applications could be fostered with external support for such training. For example, the steering committee learned through workshop discussions of the possibility that remote sensing data, service, and software providers in both the public and the private sectors could develop inexpensive programs and short courses for technical and user training and updating of skills in applied remote sensing.

As pointed out in the workshop splinter sessions, one vehicle for stimulating the development of training and course materials is the Small Business Innovation Research (SBIR) program. The SBIR is a government-wide program to encourage research and development contributions by small business, stimulate

---

[1]See Space Studies Board, National Research Council, *Assessment of Technology Development in NASA's Office of Space Science*, Washington, D.C., National Academy Press, 1998, pp. 22-23; "Continuing Assessment of Technology Development in NASA's Office of Space Science," letter from Daniel J. Fink, Chair, Task Group on Technology Development in NASA's Office of Space Science, and Claude R. Canizares, Chair, Space Studies Board, to Dr. Edward J. Weiler, Associate Administrator, Office of Space Science, NASA, March 15, 2000, pp. 6-8.

innovation in science and technology areas, develop and commercialize new technologies, products and services, and make these assets available to federal agencies.[2] The SBIR program could be used to develop training materials and courses to foster the use of remote sensing data both within and outside of federal government agencies. Moreover, remote sensing training and educational services may emerge as an important market for private companies; the SBIR could serve to encourage such market development.

The three case studies described in Chapter 2 showed that cost-effective applications take time. Organizations should be aware that applications require both initial and ongoing expenditures and that the cost of developing an application will be higher at the beginning. To reduce the budgetary impact of the initial costs, the steering committee suggests that these costs, like the benefits of remote sensing applications, be amortized over the life of the application if possible. In addition, however, new ways of reducing the start-up costs of developing applications must be identified. NASA's Office of Earth Science, Division of Applications, which is charged with securing and employing the resources required to move research into applications and fostering the operational use of remote sensing, could play a significant role in exploring more cost-effective ways to develop and implement applications.

Although the steering committee considered the nature and range of expenditures required to develop remote sensing applications, it does not attempt cost-benefit analyses or address the issue of return on investment. This important factor could not be explored in the workshop setting and was outside the scope of the workshop. However, research on the life-cycle costs and on the benefits of operational remote sensing applications could inform institutional decision making about the use of remote sensing applications. The research that has been done on cost-effectiveness is mainly proprietary market research for private sector companies. Focused cost-benefit analyses of remote sensing products and services could be used to assess whether applications are efficient and whether specific applications cost-effectively fulfill the purpose for which they were designed. Such analyses could be used to inform decisions about whether to invest in the data and infrastructure to produce remote sensing applications, to purchase the products and services from value-adding companies, or to maintain existing data systems. Studies are needed to understand the life-cycle costs and benefits of using remote sensing for major public and private sector applications, similar to the cost-benefit studies conducted in the private sector remote sensing industry.

---

[2]Public Law 106-554-Appendix I-H.R. 5667, The Small Business Innovation Research Program Reauthorization Act of 2000.

TIMELINESS OF DATA

The timeliness of remote sensing data can be an issue with respect to the initial data acquisition, the frequency of repeat coverage for detecting, monitoring, or modeling change, and the delivery of the data to the user. Because many applications of remote sensing data and information require near-real-time data, timeliness is an issue in developing new applications. Gaps of weeks to months between obtaining and processing the data are unacceptable for many operational applications, such as mitigation of potentially harmful events. For agricultural and natural resource applications, for example, remote sensing data is useful to the extent that it can provide information tied to the growth cycle and can be delivered rapidly enough to permit management interventions in this cycle. For example, data on a soybean or cotton field that are more than 1 week old are usually considered to have little value for making decisions. If insects or disease have affected a crop, a farmer may have only a few days to take corrective action once the problem has been diagnosed. Information on the status of the crop obtained as little as 2 weeks later may be irrelevant. Participants and speakers at the workshop reported that in coastal areas, timeliness is equally important. The case study on satellite and aerial remote sensing for coastal sewage discharge monitoring (Chapter 2, Box 2.3) pointed out that local water quality boards must obtain notice of sewage plumes moving toward public beaches in time to take action to protect public health. Monitoring sewage outfall and storm runoff requires data that can be processed, interpreted, and delivered to the end user in less than 24 hours.

There are natural, technical, and institutional factors that conspire to reduce the timeliness of initial remote sensing data acquisition and the frequency of repeat coverage. Cloud cover, daylight, and weather are natural factors that interfere with visibility, and for those areas that experience routine heavy cloud cover, this can be a serious limitation. At times, seasonal weather patterns, such as the very wet summer of 2000, can seriously limit the use of remote sensing for extended periods of time. Radar satellites that can provide cloud-penetrating coverage generally supply lower-resolution data than do optical systems and consequently may not meet the requirements of some applications users.[3] Technical factors influencing the timeliness of images of specific locations include orbital characteristics such as the path of the satellite and the schedule for its return flyover. The number of satellites in orbit and their availability to image specific targets also affect timely access to images.

Timeliness of data delivery also varies widely by instrument and the source

[3]Dehquanzada, Yahya A., and Florini, Ann M., *Secrets for Sale: How Commercial Satellite Imagery Will Change the World*, Washington, D.C., Carnegie Endowment for International Peace, 2000, p. 25.

of the data. For example, as noted in the coastal sewage discharge case study, data from the AVHRR, SeaWiFS, and the MODIS sensors can be obtained on a near-real-time basis. However, the spatial resolution of the data is too coarse for many applications. Data of higher resolution are often obtained at less frequent intervals.

Workshop discussions emphasized that real-time access requires the up-front design of an end-to-end data collection, processing, and distribution system. It cannot be implemented in an ad hoc manner. Information users must carefully document the need for timeliness in initial acquisition and repeat coverage of imagery. This could be done both in acquiring the data and through studies of the economic trade-offs between the value for applications of timely data and the financial costs of systems that provide timely repeat coverage. The results of such research could then be used when designing or maintaining satellite systems by both the government and the private sector.

One means of obtaining more timely data is to coordinate observations from several government and private sector Earth observing satellites. Such coordination could provide data with the same spatial, spectral, and geographic coverage.[4] This will not be easy to do, however, because the systems were initially designed to meet different sets of needs. Another approach to increasing the real-time collection of data is to increase the number of receiving stations for the data on the ground. Both government and commercial systems are relying on a small number of ground stations for data acquisition, thus further constraining the flow of data to the user community. Moreover, given the breadth of the potential applications community, broad distribution of many small, customized remote sensing products could tax a centralized architecture that relies on global-scale standard products. Workshop splinter discussions suggested the development and deployment of low-cost ground stations that could be coupled with data processing systems. The model for this approach is the present network of high-resolution picture transmission (HRPT) stations for acquiring Polar-orbiting Operational Environmental Satellite (POES) data. However, the next generation of meteorological satellites will be based on X-band downlinks, which at present are much more expensive than the L-band systems used for POES.[5] To download the collected remotely sensed satellite data, a receiving station must be within the field of view of the satellite and/or be able to communicate with the tracking data and relay satellites that can relay the data to a receiving station.

---

[4]Space Studies Board, National Research Council, *The Role of Small Satellites in NASA and NOAA Earth Observation Programs*, Washington, D.C., National Academy Press, 2000, p. 43.

[5]Researchers in the Earth science community report that a typical HRPT station can be purchased for less than $100,000, whereas a low-cost X-band station costs an estimated $300,000 to $500,000.

RELIABILITY AND CONTINUITY OF DATA

It was clear from the steering committee's discussions and the case studies that if remote sensing data are to become an intrinsic component in operational applications, there must be reliable access to the data over time. Reliability encompasses both continuity in the source of data and stability in data provision. Continuity in data sources has been difficult to achieve in the past because of the limited life span of satellites and changes in the purpose of specific instruments due to both technical advances and scientific need. Unlike the weather satellites, scientific and commercial satellites have not been designed to provide continuous, reliable data for long-term operational use. NASA's Earth science missions are designed to collect measurements for science objectives that may or may not involve long-term data collection and continuity.[6] Although operational data produced by instruments such as NOAA's polar-orbiting satellites do meet requirements for reliable and continuous access, these sensors were designed to meet short-term forecast needs, which do not always coincide with applications users' needs for long-term observations to support regulatory applications.[7]

During splinter sessions at the workshop, participants noted that because it takes many years for applications to be developed and their use made routine in an organization or government agency, managers are reluctant to commit resources to remote sensing applications when there is no assurance that the data will be available in the long run. Both commercial and government data systems may pose problems for applications users who desire assurances of the reliability and continuity of data. The potential for the failure of satellites or business strategies, or for inadequate returns on investments, could limit end users' confidence in commercial providers as a source of reliable access to remote sensing data and influence their decisions about whether or not to invest in the data. Similarly, the potential for budget cuts, policy shifts, and changes in scientific priorities could limit users' expectations for the continuity of data from government sources.

---

[6]See Space Studies Board, National Research Council, *Issues in the Integration of Research and Operational Satellite Systems for Climate Research: I. Science and Design*, Washington, D.C., National Academy Press, 2000; Space Studies Board, National Research Council, *Review of NASA's Earth Science Enterprise Research Strategy*, Washington, D.C., National Academy Press, 2000.

[7]For more background on the use of polar-orbiting satellites for long-term observational measurements (for climate monitoring), see Space Studies Board, National Research Council, *Issues in the Integration of Research and Operational Satellite Systems for Climate Research: I. Science and Design*, Washington, D.C., National Academy Press, 2000; National Environmental Satellite, Data, and Information Service, National Oceanic and Atmospheric Administration, *Climate Measurement Requirements for the National Polar-orbiting Operational Environmental Satellite System (NPOESS) Workshop Report*, Herbert Jacobowitz, ed., College Park, Md., University of Maryland, February 1997.

The steering committee heard, through repeated emphasis in the workshop, about the importance of the continuity of data series for operational applications, whether the data were produced in the public or private sector. Landsat was cited often as an example of an applications-friendly remote sensing system that has gained in utility because it provides access to usable data over what are, for remote sensing, long time periods. Yet the future of a follow-on to Landsat 7, whether a public or private sector system, is uncertain, and that uncertainty may undercut the continued utility of Landsat 7 data for new applications. If a decision is required on whether to continue a remote sensing system or to replace it with a different system, the decision should be based on applications criteria as well as scientific and technical criteria. Advisory bodies that are consulted about these decisions should include applications users in the public and private sectors. In short, there should be clearer communication between remote sensing policy makers and applied users than exists at present.

The discussions during the workshop and within the steering committee raised several other questions related to the continuity of data. Issues such as access to legacy data for applications and the need for an infrastructure and a management system for archiving legacy data will be addressed in the steering committee's second workshop, which will focus on the implications of the commercial remote sensing environment for scientific research.

DATA FORMATS AND STANDARDS

Users of remote sensing data need more consistent data formats. Data that require considerable preprocessing because they are referenced to a special map projection or a database not in common use often cause users to incur large processing costs that add to the total cost of the application. Both applications users and data suppliers incur additional costs because of the need to handle data in multiple formats. Because there are no written standards for the collection or formatting of remotely sensed images, customers who order data from multiple vendors are faced with processing data in multiple formats. Workshop participants told the steering committee that vendors generally provide the data in the format requested by the customer, but this practice requires vendors to support several types of data formats and thus merely transfers the burden to data producers.

Some applications users reported that having fixed standards for verifying and validating data, open and available protocols for developing algorithms, and standard software for processing data would encourage more widespread development and use of applications. Standard formats would allow the creation of standard data products. In the long term, standardization might help resolve some of the current impediments to the development and use of remote sensing. In the short term, workshop participants noted, vendors could post data as "geotif" files of unprocessed, raw data to enable users to make a quick decision about a data source. One response to the small number of raster data formats is to use the raster data format specifications being developed by the Federal Geographical Data Committee[8] with input from the OpenGIS Consortium.[9]

EDUCATION, TRAINING, AND THE REMOTE SENSING WORKFORCE

Multiple issues related to the remote sensing workforce have to be addressed to stimulate the pace of development of operational applications and to bridge the gap between data and information. The workshop discussions reiterated the importance of trained technical staff. Whether in the public or the private sector, an organization's capacity to incorporate remote sensing applications into its operations depends on having either technical staff with the necessary skills and understanding to process the data and transform it into usable information, or knowledgeable staff who can manage contracts with external, value-adding service providers. A major issue is the technical difficulty posed by working with remote sensing data. Because of the complexity of remote sensing images, extracting useful information requires a high level of user sophistication and training. Although technical expertise is not the only element needed to introduce or develop new applications of remote sensing, it plays a critical role. In addition, a technically trained staff member is often the only one who can bridge the gap between the technology and the needs of users within an organization. Several levels of education and training are important for sustaining the remote sensing research and applications infrastructure. Some agency representatives voiced the concern that government institutions often have an aging workforce in which few new positions are created and few younger employees are entering the system. As a consequence, there are few opportunities to hire employees with new skill sets or to transfer knowledge to new hires.

Universities play a dual role in remote sensing research and education. Uni-

---

[8]"The Federal Geographic Data Committee (FGDC) coordinates the development of the National Spatial Data Infrastructure (NSDI). The NSDI encompasses policies, standards, and procedures for organizations to cooperatively produce and share geographic data. The 17 federal agencies that make up the FGDC are developing the NSDI in cooperation with organizations from state, local and tribal governments, the academic community, and the private sector." From information accessed online at <http://www.fgdc.gov/> on December 28, 2000.

[9]The OpenGIS Consortium is an organization composed of members from the private sector, government agencies, and academic institutions in both the United States and abroad. The consortium is working to improve interoperability among systems for processing georeferenced data and general computing systems and to establish consensus on technology standards and business process innovations that would allow users to treat georeferenced data like other standard data types. See "Open GIS Consortium: Spatial Connectivity for a Changing World," available online at <http://www.opengis.org/info/brochure/brochure0599.pdf>.

versity faculty provide instruction in the use of remote sensing for scientific inquiry, including research using state-of-the-art remote sensing technologies and the development of new algorithms. Fields such as oceanography, geography, and forestry typically incorporate remote sensing as a fundamental research tool. It is also in the universities that core remote sensing research, which is vital to future advances in the field, is conducted. Academic education and research are equally important for applications. However, there is some concern about the prospects for academia's maintaining the strength necessary to meet these needs in the future. The steering committee heard through workshop discussions, for example, that because of the availability of lucrative jobs in private industry, many students take professional positions as remote sensing experts, leaving the university with a master's degree or even before obtaining an advanced degree. There was a concern for the continuation of a sufficiently strong remote sensing faculty in universities to train the next generation of remote sensing applications developers.[10]

Some universities provide technical extension training in remote sensing and other spatial information technologies for practitioners. This type of ongoing training, which is needed to support professional remote sensing experts already working in the public or private sector, includes photogrammetry, mapping, and remote sensing and geographic information systems, and it plays an essential role in maintaining a technically proficient corps of remote sensing professionals and in updating their skills. A third type of training is that provided to nontechnical users who need instruction in the use and interpretation of remote sensing information as a tool for decision making. The importance of providing this type of training should not be underestimated. Because the utility of remote sensing data is in its information content, and the ultimate users of remote sensing applications are likely to be nontechnical decision makers who influence budget decisions, it is very important that the end users understand the potential, the advantages, and the limitations of remote sensing data. End users are often ignored in remote sensing instruction and education. User training could be offered in universities or through commercial sector data providers or other private sector companies.

One issue in training is the need to standardize remote sensing training so that professionals in the field share a common background and skill set.[11] There

---

[10]See Potestio, D.S., *An Introduction to Geographic Information Technologies and Their Applications*, Washington, D.C., National Conference of State Legislatures, p. 92. In addition, the American Society of Photogrammetry and Remote Sensing is preparing a 10-year industry forecast. Information on the status of this forecast can be located online at <http://www.asprs.org/html> under the "News and External Affairs" category, which includes a link to the "Ten Year Industry Forecast."

[11]See Estes, John E., and Jensen, John R., "Development of Remote Sensing Digital Image Processing Systems and Raster GIS," *The History of Geographic Information Systems*, T. Foresman, ed., New York, Longman, Inc., pp. 163-180; and American Society for Photogrammetry and Remote Sensing, *ASPRS Certification Program*, Bethesda, Md., ASPRS, 2000.

is currently no remote sensing core curriculum, nor are there internationally recognized certification or registration requirements. The American Society for Photogrammetry and Remote Sensing and NASA are developing a remote sensing core curriculum, but its use will not be mandatory. The International Organization for Standardization (ISO) committee on spatial data standards is creating a working group to consider international standards for certification of geographic information system (including remote sensing) professionals.[12]

INTELLECTUAL PROPERTY ISSUES

Emerging intellectual property-related practices affect the cost of remote sensing applications and require new approaches to managing and safeguarding property rights. Several National Research Council reports have addressed important intellectual property issues regarding digital databases and information, including remote sensing data.[13] These studies have noted that under current copyright law the data themselves, whether produced by the public or private sector, are not copyrightable, although the original selection, coordination, and arrangement of the data in databases may be copyrightable.[14] Databases also can be protected by contract, trade secret law, and some state unfair competition law, as well as by various technological safeguards and a variety of business practices.

The steering committee learned from workshop discussions that there is a growing trend toward licensing rather than selling commercially produced databases, and thus the terms for use of the data are governed by the terms of the license. Unless licenses are constructed with the concerns of the applications community in mind, successive uses of the same data could become very costly. This expense could eventually become a disincentive to using remote sensing data, given the other applications costs that must be met.

Applications users have special concerns with respect to intellectual property rights. If an application is developed in a government agency, there may be a need to share that product throughout the agency, with officials in other levels of government, or even with the public. Applications developed by private sector remote sensing service providers, however, often may be conceived with the intent of providing related services to multiple users for a fee, or may be devel-

---

[12]Kemp, K., *Educational Challenges Workshop*, Minneapolis, Minn., University Consortium for Geographic Information Science (UCGIS), 1999, p. 6. Also see <www.ucgis.org/edu2.htm>.

[13]See National Research Council, *Bits of Power: Issues in Global Access to Scientific Data* (1997), *A Question of Balance: Private Rights and the Public Interest in Scientific and Technical Databases* (1999), and *The Digital Dilemma: Intellectual Property in the Information Age* (2000), all published by the National Academy Press, Washington, D.C.

[14]See the 1976 Copyright Act, 17 U.S.C., section 101, and *Feist Publications, Inc. v. Rural Telephone Service Co.*, 111 S. Ct. 1282 (1991).

oped for a single client who wishes to make the data available to many users over time.

Licenses already reflect several approaches to addressing the needs of multiple users. And workshop participants suggested that licensing terms could specify whether subsets of the data might be publishable, or that licenses could be priced according to the degree to which the user intends to disseminate the data, so that all users do not pay for the dissemination needs of a few. As the industry matures, it will be important to evaluate how different approaches to licensing affect the applications community and the overall development of new applications for remote sensing data.

The workshop splinter sessions raised several additional questions related to intellectual property rights and the needs of scientific remote sensing data users, such as publication of scientific research, the free and open circulation of the results of scientific research and access to that research, and the sharing of data for collaborative research—issues that will be explored in the steering committee's second workshop.

4

Recommendations

This chapter distills the results of the deliberations underlying the steering committee's findings and recommendations, which represent the consensus of the steering committee, not the workshop participants. As described in Chapter 1, the workshop provided data to the steering committee for exploring issues related to remote sensing and technology transfer. The subject of coastal zone management was selected as an applications area that could provide a useful starting point for drawing conclusions about the broader themes related to remote sensing and technology transfer.

BRIDGING GAPS

The growing availability of diverse remote sensing data opens up new opportunities for the development of remote sensing applications in both the public and private sectors. As demonstrated at the May 2000 workshop, there was broad interest among coastal zone experts in having more remote sensing applications. However, workshop participants pointed to several issues that need to be addressed before new and existing end users and remote sensing experts can take full advantage of those opportunities.

The steering committee found that moving from remote sensing data to effective civilian applications requires bridging several significant gaps. The first of these is the gap between the raw remote sensing data and the information needs of applications users. As described in Chapter 2 and illustrated in the case studies presented at the workshop, applications users need information, not data, and the process of transforming data into information is a critical step in the development of successful remote sensing applications.

The second gap to be bridged is the gap in communication and understanding between those with technical experience and training and those who are the potential end users of a remote sensing technology. The steering committee learned that unless this gap can be bridged, the transfer of remote sensing technology for the development of operational applications for new user groups will be difficult. In part this is because the applied user community rarely has the technical training necessary to understand remote sensing applications and in part because these potential users often have different information needs. As described in Chapters 2 and 3, improving communication and understanding will require that those with technical training work closely with nontechnical end users to enable them to understand both the promise and the limitations of information derived from remote sensing data. This interaction will also help the technical experts to understand end users' data and information requirements and needs. Education and training opportunities may also help to bridge this gap.

A third gap that must be bridged is the cost gap between the acquisition of remote sensing data and the development of a usable application. As experienced users of remote sensing know quite well, the purchase of data is only the first of a number of steps required to develop a successful application. A not-for-profit organization, commercial firm, or government agency that wants to incorporate remote sensing applications into its operations must be prepared to make a long-term financial investment in staff, ongoing training (both technical and user training), hardware, and software. Or the organization must be prepared to purchase these services from a value-adding provider. The various types of investments needed to maintain an applications development program, described in Chapter 4, are often overlooked.

A concern reiterated in workshop discussions was that the design of remote sensing systems—a process that has been motivated largely by scientific, military, and commercial concerns—has not taken into sufficient account the needs of applications users, which include, for example, data continuity, reliability, timeliness, and conformance to common standards, as described in Chapter 4. The needs of applications users often differ in significant ways from the needs of scientific or military users. Remote sensing image providers in both the public and the private sector must take applications needs into account in determining remote sensing system requirements.

TRANSFERRING TECHNOLOGY

Successful remote sensing applications are developed through a process of technology and knowledge transfer. Using information in the workshop presentations and the case studies, the steering committee explored technology transfer in the context of moving remote sensing data into effective applications. The steering committee found that successful technology transfer often requires two complementary actions: (1) the introduction of a new technology or product for which intellectual property can be defined, and (2) the transfer of understanding or knowledge about the technology and its application. Both aspects of technology transfer are necessary.

Technology transfer can occur across or within the public, private, and not-for-profit sectors. Bottlenecks and barriers to effective technology transfer can be found at all points in the process. The process is not defined by the nature of the organizations involved. Rather, success in technology transfer as defined by the development of usable applications is dependent on personal interactions in the process, that is, having a "spannable social distance" between each element in the technology transfer and applications process. In practice, this means that while developing new applications, the data providers and technical processors of the data must be able to understand the needs, cultural context, and organizational processes of the end users. It also means that educational and training efforts are needed to ensure that end users have a better understanding of a new technology's potential.

FINDINGS AND RECOMMENDATIONS

Life-Cycle Costs

Finding. The full, life-cycle cost of developing and using remote sensing data products goes beyond obtaining the data and includes, among others, staff for data processing, interpretation, and integration; education and training; hardware and software upgrades; and sustained interactions between technical personnel and end users. Although many of the costs are incurred at the time a technology is first employed, the life-cycle costs and benefits of remote sensing applications are not well understood.

Commercial firms have conducted studies to examine the potential market for remote sensing data and to analyze the potential cost-benefit trade-offs of using remote sensing data as opposed to products derived from other forms of information. However, most of these studies are not publicly available. Moreover, given the suggestions in workshop discussions that several cost issues are not well understood, the steering committee believes that research on the costs and benefits of using remote sensing data for noncommercial applications would benefit decision makers. Among the cost issues mentioned are the economic trade-offs of having access to timely data versus the costs of using data from systems that provide timely repeat coverage, and the costs and benefits of improving access to remote sensing data through, for instance, low-cost ground stations.

Recommendation 1. NASA's Office of Earth Science, Applications Division, in consultation with other stakeholders (e.g., agencies that use remote sensing data, such as the U.S. Geological Survey, Department of Transportation, Environmental Protection Agency, and Department of Agriculture; private companies; state and local government users; and not-for-profit institutions), should mount a study to identify and analyze the full range of short- and long-term costs and benefits of developing remote sensing applications and the full costs of their implementation by public, nongovernmental, and other noncommercial users. In addition, NASA should support economic analyses to reduce the start-up costs of developing new remote sensing applications.

Education, Training, and the Remote Sensing Workforce

Finding. Training is an integral component of efforts to bridge the gap between remote sensing professionals and end users (see Chapters 3 and 4). Remote sensing involves sophisticated technology, and specialized training is required to process the data, convert it into information, and interpret the results. Many agencies and organizations either lack the financial resources to provide such training or do not understand the importance of periodic retraining for technical staff.

Often managers do not perceive periodic training as a priority when evaluated against other organizational financial demands. For example, some at the workshop who where representatives of organizations that understand the benefits of using remote sensing information reported that even they lack sufficient in-house skill to process new sources of data and develop remote sensing applications. More education and training options are needed in remote sensing, as is an educational infrastructure that is capable of producing the number and quality of remote sensing professionals needed over the next decade and beyond.

Providing training services might, for example, represent an opportunity for the private sector, which has expertise in developing short training programs. Commercial remote sensing companies could also provide additional service to customers (and attract new users) by offering fundamental training courses on using remote sensing data. Ideally, the funding for these programs should come from both the public and private sectors. At present, however, there is no forum or mechanism for jointly funding education and training activities. The initial support for training could come from the public sector with continuation provided by the private sector if a sustainable market develops.

Recommendation 2. Federal agencies such as NASA, the National Oceanic and Atmospheric Administration, the U.S. Department of Agriculture, the U.S. Geological Survey, and others should provide the seed funding for developing remote sensing training and educational materials. Agencies should consider, as an initial step, using the Small Business Innovation Research (SBIR) program to solicit proposals for developing training materials and courses, to foster the uses of remote sensing data in applications, and to encourage commercial enterprises to provide these services.

Outreach

Finding. Reducing the social distance between application developers and end users is a means of encouraging successful technology transfer (see Chapters 2 and 3). Unless those who create applications (e.g., scientists, engineers, and technicians) and those who use them (e.g., government, not-for-profit, and private sector applied users, policy makers, and natural resource managers) understand the roles of others involved in the process, they will not be able to communicate effectively and the development of applications will suffer.

Education and training courses will help to improve end users' understanding of remote sensing but may not be sufficient for improving technicians' understanding of the end users' information needs and decision-making environments. Efforts to improve communication by reducing the "distance" among those involved in specific applications will help to foster the adoption of remote sensing technology. The county extension educator (formerly known as the county extension agent) in the agricultural community is one model for spanning the social or communications distance. More interactions among remote sensing scientists, engineers, technicians, data providers, and local, state, federal, and commercial users should be fostered whenever possible. Use of an extern program, such as that enabled by the existing Intergovernmental Personnel Act (IPA) that provides opportunities for senior staff in universities and the federal government to contribute to a different organization or sector, could facilitate technology transfer. Personnel exchanges are valuable for bridging the gap between users and technical experts.

Recommendation 3. Federal agencies, including those that produce remote sensing images and those that use them, should consider creating "extern" programs with the purpose of fostering the exchange of staff among user and producer agencies for training purposes.

For example, NASA, NOAA, and the USGS could create an extern program in collaboration with potential user agencies, such as the Environmental Protection Agency, the U.S. Army Corps of Engineers, the U.S. Department of Agriculture, the Department of Transportation, and others and in so doing could produce trained staff to serve as brokers for information and further training. Similar exchanges could be organized between universities and state and local governments and between commercial companies and government.

Recommendation 4. The Land Grant, Sea Grant, and Agricultural Extension programs should be expanded to include graduate fellowships and associateships to permit students to work at agencies that use remote sensing data. Such programs could help to improve communication and understanding among the scientists and engineers who develop applications for remote sensing data and the agencies that use them.

NASA's Space Grant program could be extended to include these training activities, much as the Land Grant program has fostered the development of agricultural extension agents.

Applications Research

Finding. Although many remote sensing applications emerge from basic research, the development of applications is not accorded the recognition associated with publication in scientific journals. Researchers have few professional incentives to produce applications. The research-to-applications model developed in other fields, such as pharmaceutical research and many fields of engineering, could be emulated by the Earth sciences. Yet even if this model were to be adopted in areas related to remote sensing, there are at present few funding opportunities for work that spans the divide between research and applications.

On the local level, many states and land grant colleges have the responsibility to use the fruits of basic research to benefit local and state issues.

Recommendation 5. Resources, separate from funding for basic research, should be made available to federal agencies such as NASA, the National Oceanic and Atmospheric Administration, the Environmental Protection Agency, the U.S. Geological Survey, the Department of Transportation, the National Science Foundation, and others for support of research on remote sensing applications and remote sensing applications derived from basic research. In addition, these agencies should establish joint research announcements aimed at fostering the development of applications for remote sensing data through basic research.

Requirements of Applications Users

Finding. Many remote sensing applications have specific requirements, including continuity in data collection, consistency in format, frequency of observations, and access to comparable data over time. It is important that the requirements of those who use applications are communicated to both public and private sector data producers throughout the process of designing new technologies and producing and disseminating remote sensing data.

Many users, for example, require historical data to provide a basis for assessing land-cover and land-use changes and for verifying if natural anomalies are consistent in their size and intensity.

Recommendation 6. Both public and private sector data providers should develop mechanisms to obtain regular advice and feedback on applications requirements for use in their planning processes. Advisory bodies that are consulted for input to these decisions should routinely include applications users.

Recommendation 7. Data preservation should be addressed by all data providers as a routine part of the data production process to ensure continuity of the data record and to avoid inadvertent loss of usable data.

Standards and Protocols

Finding. The lack of standard data formats, open and available protocols, and standard validation and verification information inhibits the spread of remote sensing applications (see Chapter 3).

For example, new sensors acquire data at bit rates different from those of existing data sets collected over an application site. Consequently, users must bear the cost of reformatting and of assessing the accuracy of new data sets. Validation and verification procedures, while now in place, were not established until recently, when it became evident that vendors would be providing non-calibrated data. The use of standardized formats in remote sensing and other spatial data technologies would facilitate the creation and widespread use of standard data products.

Recommendation 8. The use of internationally recognized formats, standards, and protocols should be encouraged for remote sensing data and information. The work of the OpenGIS Consortium and the Federal Geographic Data Committee serves as an important international and national coordinating mechanism for efforts in standards development that should be continued.

These and other entities pursuing common remote sensing data formats and standards should consult with the sensor and software vendors to ensure that data acquired from the use of new technologies for data acquisition, analysis, and storage and distribution are consistent with other data sets.

Utility of Workshop Format

Finding. In general, the workshop as a mechanism for gathering data provided the steering committee with the information and insight it needed to understand issues related to technology transfer and remote sensing applications and to make recommendations about more effective ways to foster the development of applications.

Workshop Format

The steering committee identified several strengths and some areas for improvement in its use of the workshop as a means of gathering information. These "lessons learned" are important to the steering committee as it plans the second and third workshops in this three-part series.

The mix of plenaries, panels, and splinter group sessions worked well, combining expert presentations and case studies that informed workshop attendees about research on the subject as well as concrete experiences with applications of remote sensing and technology transfer. For example, a presentation by a researcher who had studied the technology transfer process and who summarized a small portion of the history of research in the field was valuable, especially to those with experience primarily in the technical aspects of remote sensing applications. The splinter groups, in turn, provided a forum for the workshop attendees to discuss the issues that arose in the plenary sessions and to make their experience and observations available to the steering committee in an interactive manner.

In addition, the steering committee obtained useful input from sponsoring agencies during both the development and staging of the workshop. In a pre-workshop planning meeting federal agency representatives provided information on the nature of problems to which remote sensing might be applied, on challenges presented by efforts to apply remote sensing, and on impediments to technology and knowledge transfer within the agencies and constituencies at large.

Although the steering committee made an effort to invite a diverse group of participants from the public, private, and academic sectors, some portions of the applications community were not well represented in the workshop, especially the commercial sector. For the two upcoming workshops, the steering committee plans to redouble its efforts to attract representatives of the commercial sector, both as speakers and as participants in the discussions.

The steering committee found that the workshop size and setting affected the nature of discussions. The large audience and the setting (approximately 75 attendees in an auditorium), although effective for plenary presentations, were not conducive to informal interactions across the diverse sectors of the applications community (academic, governmental, commercial, and not-for-profit) and with the steering committee and the workshop speakers. The steering committee is considering holding the next workshop in a setting that facilitates a more interactive discussion.

Regarding its own participation in the workshop, the steering committee noted that leading each splinter group (taking responsibility for moderating the discussion, seeing that all viewpoints were expressed, summarizing and reporting on the discussion to the full workshop, and drafting the relevant section for this report) had the benefit of allowing individual members to share their expertise at the workshop. At the same time, it may have detracted from the steering committee's focus on its overarching role to understand the full range of issues and experiences reported at the workshop.

Appendixes

A

Biographical Information for Steering Committee Members and Workshop Speakers

STEERING COMMITTEE MEMBERS

Roberta Balstad Miller, *Chair*, has worked and published extensively in the areas of science and technology policy and human interactions in global environmental change. She received her Ph.D. from the University of Minnesota. Currently the director of the Center for International Earth Science Information Network at Columbia University, she was previously a staff associate with the Social Science Research Council (1975 to 1981), the founding executive director of the Consortium of Social Science Associations (1981 to 1984), and director of the Division of Social and Economic Science at the National Science Foundation (NSF) (1984 to 1993). She received NSF's Meritorious Service Award in 1993. Dr. Miller has served as chair of a number of scientific advisory groups, including the North Atlantic Treaty Organization (NATO) Advisory Panel on Advanced Science Institutes and Advanced Research Workshops; the Committee on Science, Engineering, and Public Policy of the American Association for the Advancement of Science; the Human Dominated Systems Directorate of the U.S. Man in the Biosphere Program; and others. From 1992 to 1994, she served as vice president of the International Social Science Council. Dr. Miller's National Research Council (NRC) service includes former membership on the Space Studies Board, the Board's Task Group on Research and Analysis Programs, and the Climate Research Committee. She currently serves on the Committee on Global Change Research.

Mark R. Abbott was an acting assistant professor for the Section of Ecology and Systematics at Cornell University (1978 to 1979) and a postgraduate researcher for the Institute of Ecology at the University of California at Davis (1979 to 1980). From 1980 to 1982, Dr. Abbott was a NATO/NSF postdoctoral fellow of ocean ecology at the Institute of Ocean Studies in Sidney, British Columbia. He was a member of the technical staff of the Oceanography Group at the Jet Propulsion Laboratory from 1982 to 1988. At the same time, he was also an assistant adjunct professor of the Marine Life Research Group at the Scripps Institution of Oceanography. In 1988, Dr. Abbott joined the faculty of the College of Oceanic and Atmospheric Sciences at Oregon State University, where he is dean of the college and a professor of biological oceanography. He currently serves on the National Aeronautics and Space Administration (NASA) Earth Observing System (EOS) Investigators Working Group (1989 to the present) and is a member of the EOS Moderate Resolution Imaging Spectrometer Science Team (1989 to the present). He currently chairs the EOS Payload Panel (1995 to the present). In addition, he serves on NSF's Joint Global Ocean Flux Study Science Executive Committee (1996 to the present). Dr. Abbott is chair of the Space Studies Board's Committee on Earth Studies.

Lawrence W. Harding, Jr., is a research professor at the University of Maryland Center for Environmental Science, with appointments at Maryland Sea Grant and the Horn Point Laboratory. His research focuses on the use of aircraft and satellite remote sensing of ocean color to study phytoplankton responses to nutrient enrichment in estuarine and coastal waters. He also directs Sea Grant educational activities in remote sensing in collaboration with NASA scientists. His main interests include coordination of a regional, multiplatform remote sensing program in the Chesapeake Bay region to further the understanding of ecosystem health by applying new technologies to contemporary ecological issues.

John R. Jensen is a Carolina Distinguished Professor of geography and geographic information systems (GIS) and director of the Remote Sensing Center at the University of South Carolina. His research focuses on remote sensing of vegetation biophysical resources, especially inland and coastal wetlands; remote sensing of urban, suburban and land-use cover; development of improved digital image processing classification, change detection, and error evaluation algorithms; and development of educational materials for remote sensing instruction. Dr. Jensen has conducted contract and grant research for the Department of Energy's Savannah River Site, NASA commercial applications, and National Oceanic and Atmospheric Administration (NOAA) CoastWatch. He is the author of a textbook on remote sensing, *Introductory Digital Image Processing: A Remote Sensing Perspective*, and was president of the American Photogrammetry and Remote Sensing Society in 1996.

Chris J. Johannsen is director of the Laboratory for Applications of Remote Sensing and a professor of agronomy at Purdue University. His research interests are in spatial, spectral, and temporal aspects of remote sensing relating to GIS as applied to precision agriculture, land resource development, and land degradation. He was director of the Environmental Sciences and Engineering Institute (previously Natural Resources Research Institute) (1988 to 1995) and director of the Agricultural Data Network (1985 to 1987) at Purdue University. From 1981 to 1985, Dr. Johannsen was the director of the Geographic Resources Center, Extension Division, at the University of Missouri at Columbia. Dr. Johannsen has been named fellow to the American Society of Agronomy, the Soil Science Society of America, and the Soil Conservation Society of America and is a member of the International Soil Society, the American Society of Photogrammetry, and Sigma Xi. He has served on the Space Studies Board's Committee on Earth Studies (1995 to 1998), the Committee on NASA Information Systems (1986 to 1987), and the Panel on Earth Resources (1982 to 1983).

Molly Macauley is a senior fellow at Resources for the Future (RFF), where she directs the space economics research program. Her research interests include space economics and policy, recycling and solid waste management, urban transportation policy, and the use of economic incentives in environmental regulation. An economist at RFF since 1983 and a long-time analyst of the commercial use of space technology, Dr. Macauley offered her views to Congress in May 1997 on how government can foster burgeoning commercial ventures into satellite remote sensing. One of her major research projects looks at the ongoing economic—as well as privacy, security, and other—implications of U.S. companies selling images photographed by privately owned satellites in outer space. Her other research projects involve exploring the use of economic incentives to manage space debris; the allocation of scarce energy, water, utilities, and telecommunications resources on the International Space Station; the value of geostationary orbit; and the value of information, particularly information derived from space-based remote sensing. She was a member of the Space Studies Board's Task Group on Setting Priorities for Space Research and the NRC Committee on the Impact of Selling the Federal Helium Reserve. She currently serves on the NRC's Aeronautics and Space Engineering Board's Committee for the Assessment of NASA's Space Solar Power Investment Strategy.

John S. MacDonald is a consultant and is chairman of the Institute for Pacific Ocean Science and Technology. He is one of the founders of MacDonald, Dettwiler and Associates Ltd., where he was responsible for all aspects of business operations, overall strategic leadership, technical leadership, and market positioning worldwide. Dr. MacDonald's professional interests lie in the areas of advanced digital systems engineering, remote sensing, and image processing. He led the design team for the first Landsat ground-processing system produced by

MDA, Ltd., and was involved in the early development of synthetic aperture radar processing at this company. His technical activities have been in the areas of information extraction from advanced sensor systems and the applications of remote sensing with particular emphasis on the physics of the backscatter process and the use of integrated datasets as a means of increasing the ability to extract useful information from remotely sensed data.

Jay S. Pearlman is development team manager at TRW, Inc., in Redondo Beach, California. His background includes basic research, program management, and program development in sensors and systems. He has played an important role in the development and implementation of new concepts and capabilities for both the military and the civil sectors of the U.S. government. He is currently working on the EO-1 Hyperion sensor as the project's principal investigator and is actively involved with the EO-1 Science Validation Team in assessing the benefits of hyperspectral imagery. Dr. Pearlman is also involved in an assessment of the viability of multispectral and hyperspectral commercial applications.

PLENARY SPEAKERS

John E. Estes[1] was a professor of geography at the University of California in Santa Barbara (UCSB). His areas of specialization included interpretation and analysis of remote sensor data and use of geographic information systems (GIS) for land use/land cover and regional-resource-base determination and evaluation. Prior to coming to UCSB in 1969, Professor Estes worked in both government—as an intelligence analyst for the Central Intelligence Agency—and industry—for the Science Services Division of Texas Instruments, Inc. He served as senior visiting scientist for the Universities Space Research Association, working with the National Mapping Division of the United States Geological Survey (USGS) and the NASA Office of Mission to Planet Earth. Professor Estes received his Ph.D. from the University of California in Los Angeles in 1969. He conducted contract and grant research on both the fundamental and applied aspects of the use of remote sensor and GIS. This work included studies for NASA on land-use/land-cover identification and change detection, validation of land-cover products, mapping of protected areas, crop identification, and modeling of water demand, among others. He worked with the U.S. Forest Service on fire fuels monitoring and modeling, the National Biological Survey on biodiversity protection and GAP analysis, and the USGS and National Oceanic and Atmospheric Administration on the detection of marine oil pollution. Work conducted for other federal agencies, such as the U.S. Environmental Protection Agency, Department of Energy, and Department of Defense, emphasized hazard and pollution detection,

[1]Dr. Estes died on March 9, 2001.

monitoring, modeling, and resources management, while his work for the National Science Foundation emphasized GIS and spatial analysis.

W. Jeff Lillycrop is the director of the Joint Airborne Lidar Bathymetry Technical Center of Expertise (JALBTCX), a partnership between the U.S. Army Engineer District Mobile, the U.S. Naval Oceanographic Office, and the U.S. Army Engineer Research and Development Center's Coastal and Hydraulics Laboratory. The JALBTCX operates the SHOALS airborne lidar system and performs R&D to expand the capabilities of airborne lidar to support Department of Defense requirements. He has worked for the U.S. Army Corps of Engineers since 1983, and he holds a master's of science degree in coastal and oceanographic engineering from the University of Florida. Mr. Lillycrop is a trustee of the Hydrographic Society of America and has published more than 50 technical papers related to airborne lidar, surveying and mapping, sediment management, and other topics related to coastal engineering.

Eugene Paul Meier is currently assigned to the U.S. Environmental Protection Agency's (EPA's) National Exposure Research Laboratory and is stationed at the Stennis Space Center in Mississippi as the EPA's Office of Research and Development (ORD) liaison to the Gulf of Mexico Program (GMP). Dr. Meier is responsible for linking the research and technical capabilities of ORD to GMP requirements. His responsibilities include coordinating GMP activities and requirements in the development of the ORD research program; leading the GMP's Strategic Environmental Assessment Team; chairing the GMP's Monitoring, Modeling, and Research Committee; acting as GMP lead for its Data and Information Transfer Committee; and providing technical assistance for GMP activities. Dr. Meier has 30 years of experience in basic and applied research related to health and environmental science. He has specialized experience in the development of methods for analysis of environmental samples; development of methods for management and disposal of pesticides and hazardous waste; quality assurance in environmental monitoring; applications of remote sensing technology; and management of environmental monitoring programs. He received his B.S. degree in chemistry from Texas A&M University in 1965 and his Ph.D. in analytical chemistry from the University of Colorado in 1969.

Michael K. Orbach is a professor of marine affairs and policy and the director of the Duke University Marine Laboratory and the Coastal Environmental Management Program in the School of the Environment at Duke University. His B.A. is in economics from the University of California at Irvine, and his M.A. and Ph.D. are in cultural anthropology from the University of California at San Diego. From 1976 to 1979 he was a social anthropologist and social science advisor with the National Oceanic and Atmospheric Administration in Washington, D.C. From 1979 to 1982 he was associate director of the Center for Coastal Marine Studies at the University of California at Santa Cruz. From 1983 to 1993 he was professor of anthropology in the Department of Sociology and Anthropology and a senior scientist with the Institute for Coastal and Marine Resources at East Carolina University. He joined Duke, with offices at the Duke Marine Laboratory in Beaufort, North Carolina, in 1993. Dr. Orbach has performed research and has been involved in coastal and marine policy on all coasts of the United States and in Mexico, Central America, the Caribbean, Alaska, and the Pacific and has published widely on social science and policy in coastal and marine environments. Among his publications are *Hunters, Seamen and Entrepreneurs: The Tuna Seinermen of San Diego* (University of California Press, 1977); "U.S. Marine Policy and the Ocean Ethos" (*Marine Technology Society Journal,* 1982); *North Carolina and the Sea: An Ocean Policy Analysis* (with D. Moffitt et al., North Carolina Department of Administration, 1985); and "A Fishery in Transition: The Impact of Urbanization on Florida's Spiny Lobster Fishery" (with J. Johnson, *City and Society,* 1991).

J. David Roessner is a professor of public policy at Georgia Institute of Technology and associate director of the Science and Technology Policy Program at SRI International. Prior to joining the Georgia Tech faculty in 1980, he was principal scientist and group manager for Industrial Policy and Planning at the Solar Energy Research Institute in Golden, Colorado. He served as a policy analyst with the National Science Foundation's R&D Assessment Program and, subsequently, as acting leader of the Working Group on Innovation Processes and Their Management in the Division of Policy Research and Analysis at NSF. Previously he was a research associate at the Bureau of Social Science Research, Inc. His first professional position was as a development engineer for Hewlett-Packard Co. in Palo Alto, California. Dr. Roessner's research interests include national technology policy, the evaluation of research programs, the management of innovation in industry, technology transfer, and indicators of scientific and technological development. In addition to numerous technical reports, he has published papers in policy-oriented journals and has authored and edited several books. Dr. Roessner received B.S. and M.S. degrees in electrical engineering from Brown University and Stanford University, respectively. He returned to graduate school after working at Hewlett-Packard to receive a master's degree in science, technology, and public policy from Case Western Reserve University in 1967 and a Ph.D. in the same field in 1970.

Jan Svejkovsky is the founder and president of Ocean Imaging, Inc., where he is responsible for managing and directing all scientific and corporate developments. The company focuses on applied research and facilitates technology transfer for new operational applications. Dr. Svejkovsky is principal investigator on research grants from NOAA, NASA, NSF, the Navy, the state of California, and corporations. Through Ocean Imaging, he has also developed commercial programs for near-real-time generation and dissemination of satellite-based ocean environmental analyses to research vessels, the offshore oil industry, and commercial and sport fishing fleets in the United States and foreign countries. Ocean Imaging also provides contractual satellite oceanography support for numerous university research teams. Dr. Svejkovsky's prime interest is in identifying new potential markets for remote sensing technology and developing customized products and services for those markets. In recent years, he directed advanced development and commercialization of satellite and nonsatellite oceanographic techniques for diverse research and coastal applications, including sewage, storm runoff and other pollution effluent monitoring (using optical, infrared (IR), and synthetic aperture radar (SAR) sensors); high-resolution surface current detection (using IR, SAR, and optical imagery); and multispectral algorithms for bathymetry surveys and bottom substrate mapping. Since mid-1998, Ocean Imaging has operated its own multispectral aerial sensor for coastal research and environmental monitoring and, since 1999, rapid-response agricultural remote sensing.

Stephen J. Walsh is a professor of geography and the director of the Landscape Characterization and Spatial Analysis Laboratory in the Department of Geography at the University of North Carolina (UNC), Chapel Hill. He is the former Amos H. Hawley Professor of Geography and Director of the Spatial Analysis Unit at the UNC Carolina Population Center, as well as the past chair of the Remote Sensing and GIS Specialty Groups of the Association of American Geographers (AAG). In 1997, Dr. Walsh was awarded the Outstanding Contribution Award and Medal from the Remote Sensing Specialty Group of the AAG, and in 2000 he was awarded Research Honors from the Southeastern Division of the AAG and was a First Prize recipient of the 2000 ERDAS Award for Best Scientific Paper in Remote Sensing. He is on the editorial boards of *Plant Ecology*, *Journal of Geography*, *The Professional Geographer*, *GeoCarto International*, and the *Southeastern Geographer* and recently co-edited special issues on remote sensing in *Journal of Vegetation Science* and *Geomorphology*. Professor Walsh's research interests are in remote sensing, GIS, spatial analysis, physical geography, and population-environment interactions. His current projects include research being conducted in Ecuador and Thailand on land-use and land-cover dynamics associated with deforestation and agricultural extensification, in Montana on alpine tree line and environmental change, and in North Carolina on landscape dynamics and ecological gradients.

PANELISTS

Robert Arnone has been involved with ocean color remote sensing optics for the last 17 years and currently heads the Ocean Optics and Remote Sensing Section in the Ocean Technology Directorate at the Naval Research Laboratory in the Stennis Space Center. He leads a team of 15 people that is currently involved in basic, exploratory, and applied sciences in ocean optical and remote sensing by satellite and aircraft sensing of the ocean surface. The section is involved in national and international oceanography programs and maintains satellite receiving antennas and advanced in situ optical instrumentation. He has been on the adjunct faculty to the University of Southern Mississippi (Marine Science Department) since 1989, has served on several graduate student committees, and has approximately 5 students. He is responsible for more than 40 publications and 100 presentations. His research interests include ocean color algorithm development, biophysical processes in coastal and open oceans, and satellite databases of optical properties.

Anne Hale Miglarese is chief of the Coastal Information Services branch of the National Oceanic and Atmospheric Administration's Coastal Services Center. In addition to holding bachelor's and master's degrees in geography from the University of South Carolina, Ms. Miglarese also graduated from Leadership South Carolina and the Executive Institute. Ms. Miglarese is past chairwoman of the South Carolina State Mapping Advisory Committee and a board member of the Urban and Regional Information Systems Associations. She is currently the chair of the Federal Geographic Data Committee Subcommittee on Bathymetric and Nautical Charting Data, the U.S. Department of Commerce representative to the Civilian Applications Committee, and a member of the editorial board of *Geo Info Systems* magazine.

Walter Schmidt received his Ph.D. from Florida State University. He is currently the state geologist of Florida and chief of the Florida Geological Survey. His background includes degrees in oceanographic technology, geology, and marine geology. He is a former president of the Association of American State Geologists and a current member of the NRC's Ocean Studies Board. Dr. Schmidt's research interests include stratigraphy, hydrogeology, and environmental geology applied to public policy.

Michael Thomas was appointed as director of the Applications, Commercial, and Education Division in NASA's Earth Science Enterprise in March 2000. He comes to the Earth Science Enterprises from the Stennis Space Center, where he was deputy program manager for the NASA Commercial Remote Sensing Program. Before joining NASA, Dr. Thomas worked in the private sector, where he directed the development of new products for government and commercial use; planned and implemented the transition of new capabilities from the laboratory to operational settings; acted as liaison between corporate R&D and operational divisions; conducted his own research in artificial intelligence, pattern matching, and natural language understanding; and taught at the university level. Dr. Thomas has several degrees in anthropology, each with a different focus. He earned a B.A. (archeology) from the University of Texas at Austin (1973) and an M.A.

(cultural anthropology) from Washington State University (1976). After completing a Fulbright-Hayes Fellowship in Java, Indonesia, he went on to earn a Ph.D. (linguistics) at Washington State University (1978). Dr. Thomas is fluent in Indonesian and has a working knowledge of Malay, as well as training in Spanish, German, Dutch, Samoan, and Chamorro, the language spoken by the original people of Guam and the Northern Mariana Islands.

James Yoder is a professor of oceanography at the Graduate School of Oceanography (GSO) of the University of Rhode Island. He currently serves as interim dean of GSO and plans to return to the faculty later this year. He received his Ph.D. in 1978, also from the GSO. His first position was as a research associate at the Skidaway Institute of Oceanography in Savannah, Georgia, where he was promoted to professor in 1989. From 1986 to 1988, Dr. Yoder was a visiting senior scientist at the Jet Propulsion Laboratory, assigned to the Ocean Branch in the Office of Space Science and Applications at NASA Headquarters in Washington, D.C. He managed the ocean color research program and helped get the SeaWiFS ocean color sensor program under way. In 1989, he joined the oceanography faculty at GSO and in 1993 became GSO's associate dean for academic affairs. In the fall of 1996, he returned to NASA Headquarters for a year to manage the Biological Oceanography program in the Office of Mission to Planet Earth (now the Office of Earth Science). Dr. Yoder has published more than 50 articles in phytoplankton ecology and ocean remote sensing. His present research interests include studying the relationships between physical and biological/biogeochemical processes at regional to basin scales using satellite data as the primary tool. He has served on many national and international committees, including the U.S. JGOFS Steering Committee, the International Ocean Color Coordinating Committee, and the Scientific Executive Committee of NASA's Earth Observing System program, and is the president-elect of the Oceanography Society.

B

Workshop Agenda and Participants

AGENDA

Wednesday, May 3, 2000

8:30 a.m.	Introduction and Overview	Roberta Balstad Miller
8:45	Plenary Speaker I: Technology Transfer Process *David Roessner, Georgia Institute of Technology*	
9:30	Plenary Speaker II: Emerging Technologies for Remote Sensing and Geospatial Data *Stephen Walsh, University of North Carolina, Chapel Hill*	
10:15	Break	
10:30	Plenary Speaker III: Science and Policy Issues in the Coastal Zone *Michael Orbach, Duke University Marine Laboratory*	
11:15	Case Study Presentations	Roberta Balstad Miller
11:20	EPA Advanced Monitoring Program: Application of the SeaWiFS for Coastal Monitoring of Harmful Algal Blooms *Eugene Meier, Environmental Protection Agency*	

11:50	U.S. Army Corps of Engineers SHOALS Airborne Lidar Bathymetry Program *Jeff Lillycrop, U.S. Army Corps of Engineers*
12:20 p.m.	Lunch
1:20	Satellite and Aerial Remote Sensing for Coastal Sewage Discharge and Run-off Monitoring Project *Jan Svejkovsky, Ocean Imaging, Inc.*
1:50	General Discussion and Question and Answer Session with Case Study Panelists Moderator: Mark Abbott, Oregon State University
2:20	Introduction to Splinter Sessions — Roberta Balstad Miller
2:30	Break
2:45	Splinter Sessions

Group A:	Education/Workforce Development
Moderator:	John Jensen, University of South Carolina
Group B:	Institutional Issues in Technology Transfer
Moderator:	Mark Abbott, Oregon State University
Group C:	Policy Issues
Moderator:	Molly Macauley, Resources for the Future
Group D:	Technical Issues in Technology Transfer
Moderator:	Chris Johannsen, Purdue University
Group E:	User Awareness and Needs
Moderator:	Larry Harding, Horn Point Environmental Laboratory/Maryland Sea Grant

5:30	Adjourn
6:15	Working Session and Dinner for Steering Committee

Thursday, May 4, 2000

8:30 a.m.	Introduction of Agenda	Roberta Balstad Miller
8:35	Plenary Speaker IV: Comparative Perspectives on Technology Transfer: GIS and GPS *Jack Estes, University of California, Santa Barbara*	
9:30	Splinter Session Reports (Problems/Barriers and Solutions/Actions)	
10:20	Break	
10:35	Panel Discussion on Remote Sensing for Coastal Zone Science and Applications Moderator: Walter Schmidt, Florida State Geological Survey Panelists: Robert Arnone, Naval Research Laboratory Jan Svejkovsky, Ocean Imaging, Inc. Anne Hale Miglarese, NOAA Coastal Services Center Michael Orbach, Duke University Marine Laboratory Michael Thomas, National Aeronautics and Space Administration James Yoder, University of Rhode Island	
12:05 p.m.	Wrap-up	Roberta Balstad Miller
12:30	Adjourn	

PARTICIPANTS

Mark Abbott, Oregon State University
Joseph Alexander, NRC Space Studies Board
Rodney Anderson, Veridian ERIM Information Analysis Center
Kirsten Armstrong, NRC Space Studies Board
Robert Arnone, Naval Research Laboratory
John Baker, RAND
Jan Baxter, EPA Region 9
Matthew Bechdol, NASA Goddard Space Flight Center
David Bruggeman, George Washington University
William Burgess, Maryland Department of Natural Resources
Paul Burt, New Jersey Department of Environmental Protection

Richard Buss, Jr., NASA Goddard Space Flight Center
Long Chiu, George Mason University
Marie Colton, NOAA Office of Research and Applications
Matthew Cook
Stan Daberkow, USDA Economic Research Service
Lee Dantzler, NOAA NESDIS
Bruce Davis, NASA Stennis Space Center
Hank Drahos, Jr., NOAA NESDIS
Julie Esanu, NRC Space Studies Board
John Estes, University of California, Santa Barbara
Lawrence Friedl, EPA, Office of Research and Development
Susan Gartner, Earthwatch Institute
Richard Gomez, George Mason University
Morgan Gopnik, NRC Ocean Studies Board
Lawrence Harding, Horn Point Environmental Laboratory
Fred Henderson, Hendco Services
Emil Horvath, USDA National Resource Conservation Service
John Jensen, University of South Carolina
Chris Johannsen, Purdue University
Bret Johnson, George Washington University
Bruce Kiracofe, Shenandoah Mountain Geographics, Inc.
James Koziana, NASA Goddard Space Flight Center
Gary Krauss, Geodigital Mapping, Inc.
Murali Krishna, Jawaharlal Nehru Technological University, India
Subhash Kuvelker, Kuvelker Law Firm
W. Jeff Lillycrop, U.S. Army Corps of Engineers
George Loeb, EPA Coastal Management Branch
Ariovaldo Luchiari, Jr., University of Nebraska, Lincoln
Wilson Lundy, NASA Langley Research Center
John Lyon, EPA Office of Research and Development
Molly Macauley, Resources for the Future
Tony MacDonald, Coastal States Organization
R. Mann, U.S. Army Corps of Engineers
John Marra, NASA
James McManus, George Mason University
Eugene Meier, EPA Office of Research and Development
Jill Meyer, NOAA NESDIS
Anne Miglarese, NOAA Coastal Services Center
Roberta B. Miller, Columbia University Center for International Earth Science Information Network
Nora Normandy, NASA
Michael Orbach, Duke University Marine Laboratory

Amy Owsley, EPA Office of Wetlands, Oceans, and Watersheds, Coastal Management Branch
Tom Palmerlee, NRC Transportation Research Board
Paul Pan, EPA Oceans and Coastal Protection Division
Lawrence Pettinger, USGS
Jim Plasker, American Society of Photogrammetry and Remote Sensing
Josie Quintrell, Maine Coastal Program
Gregg Reinecke, Geodigital Mapping, Inc.
B. Robustell, George Washington University
David Roessner, Georgia Institute of Technology
Sally Rood, Federal Laboratory Consortium
Jim Schepers, USDA Agricultural Research Service
Walter Schmidt, Florida State Geological Survey
Carl Schoch, Kachemak Bay National Estuarine Research, Alaska
Robert Schuster, New Jersey Department of Environmental Protection
H. Semerjian, National Institute of Standards and Technology
Edwin Sheffner, NASA; California State University, Monterey Bay
Dennis Smith, Environmental Systems Research Institute
Jan Svejkovsky, Ocean Imaging, Inc.
Pamela Taylor, NOAA NESDIS
William Teng, NASA Goddard Space Flight Center
K. Thirimulai, Department of Transportation
Michael Thomas, NASA
Brett Thomassie, EarthWatch Inc.
James Tilton, NASA Goddard Space Flight Center
Grady Tuell, NOAA National Geodetic Survey
Alex Tuyahov, NASA
Paul Uhlir, NRC Office of International Affairs
Lisa Vandermark, NRC Board on Earth Sciences and Resources
Dan Walker, NRC Ocean Studies Board
Stephen Walsh, University of North Carolina, Chapel Hill
Ming-Ying Wei, NASA
Pamela Whitney, NRC Space Studies Board
David Williams, EPA
Hank Wolf, George Mason University
Leslie Wollack, Regional Application Center for the Northeast
Charles Wooldridge, NOAA NESDIS
James Yoder, University of Rhode Island

C

Planning Meeting Agenda

MONDAY, DECEMBER 13, 1999

Closed Session

8:00 a.m.	Continental Breakfast	
8:30	Chair's Remarks	R. Miller

Open Session

9:00	Introduction • Overview of workshops • Case study example • Issues to emphasize • Workshop structure	R. Miller
9:45	National Oceanic and Atmospheric Administration • NESDIS • NOS	 I. Hakkarinen/R. Masters R. Stumpf
10:15	Environmental Protection Agency	L. Friedl
10:35	Break	

10:50	U.S. Army Corps of Engineers	C. Chesnutt
11:10	NASA	
	• Headquarters	N. Maynard/A. Carlson
	• Stennis	B. Davis/M. Thomas
11:40	Department of Transportation	K. Thirumalai
12:00 p.m.	Working Lunch—Other Public Input	
1:00	Structure and Goals for Workshop	R. Miller
	• Case study example	
	• Issues to emphasize	
	• Number of and theme for splinter groups	
2:30	Break	
2:45	Structure and Goals (continued)	
4:00	Other Issues	R. Miller
	• Agreement on dates for first workshop	
	• Linkage to workshops II and III	
5:00	Adjourn and Attend Reception	
5:45	Dinner for Committee and Invited Guests	

TUESDAY, DECEMBER 14, 1999

Closed Session

8:00 a.m.	Continental Breakfast	
8:30	Chair's Remarks	R. Miller
	Bias, Composition, and Orientation	
	Discussion	J. Alexander

Open Session

9:15	Structure and Goals Follow-up	
	• Review of outline and strawman agenda	
	• Suggestions on speakers	

10:30 Break

10:45 Workshop Contributions R. Miller
- Possible inputs (agencies, public and private remote sensing image providers, value-adding entities, users)
- Preworkshop materials and data gathering

11:30 Wrap-up and Recap R. Miller
- Status of meeting objectives
- Other issues?

12:00 p.m. Lunch

Closed Session

1:00 Steering Committee Meeting R. Miller

3:00 Adjourn

D

Acronyms

ARC	affiliated research center
AVHRR	advanced very high resolution radiometer
DOT	Department of Transportation
EOCAP	Earth Observations Commercialization/Applications Program
EOS	Earth Observing System
EOSAT	Earth Observation Satellite Company
EPA	Environmental Protection Agency
ESE	Earth Science Enterprise
FGDC	Federal Geographical Data Committee
GIS	geographic information systems
GPS	Global Positioning System
HAB	harmful algal bloom
HRPT	high-resolution picture transmission
NASA	National Aeronautics and Space Administration
NATO	North Atlantic Treaty Organization
NESDIS	National Environmental Satellite Data and Information Service
NOAA	National Oceanic and Atmospheric Administration
NRC	National Research Council

NSDI	National Spatial Data Infrastructure
NSF	National Science Foundation
POES	Polar-orbiting Operational Environmental Satellite
SBIR	Small Business Innovation Research (program)
SeaWiFS	Sea-viewing Wide Field-of-view Sensor
SHOALS	Scanning Hydrographic Operations Airborne Lidar Survey
SPOT	Système pour l'Observation de la Terre
TIROS	Television Infrared Observing Satellite
USACE	U.S. Army Corps of Engineers
USDA	U.S. Department of Agriculture
USGS	U.S. Geological Survey